# BE CLIMATE CLEVER

DK

# BE CLIMATE CLEVER

BY AMY & ELLA MEEK

Foreword by

## STEVE BACKSHALL

DK | Penguin Random House

**Written by** Amy and Ella Meek
**Consultant** Professor Daniel Parsons
**Design** Collaborate Agency
**Illustrator** Sarah Goodreau
**Acquisitions Editor** Fay Evans
**Senior Editor** Marie Greenwood
**Designers** Sonny Flynn, Holly Price
**Production Editor** Abigail Maxwell
**Production Controller** Ena Matagic
**Managing Editor** Jonathan Melmoth
**Managing Art Editor** Diane Peyton Jones
**Deputy Art Director** Mabel Chan
**Publishing Director** Sarah Larter

First published in Great Britain in 2022
by Dorling Kindersley Limited
DK, One Embassy Gardens, 8 Viaduct Gardens,
London, SW11 7BW

The authorised representative in the EEA is
Dorling Kindersley Verlag GmbH. Arnulfstr. 124,
80636 Munich, Germany

Copyright © 2022 Dorling Kindersley Limited
A Penguin Random House Company
10 9 8 7 6 5 4 3 2 1
001–326508–April/2022

A CIP catalogue record for this book
is available from the British Library.
ISBN: 978-0-2415-3339-0

Printed and bound in Great Britain by
Clays Ltd, Elcograf S.p.A.

MIX
Paper from
responsible sources
FSC™ C018179
FSC www.fsc.org

For the curious
**www.dk.com**

# CONTENTS

## We can make a difference!

# FOREWORD
## by Steve Backshall

I was sitting reading about dinosaurs with my three-year-old this morning, talking about asteroids, extinctions, and big topics about how our world changes, when suddenly he hit me with the killer question: **"Daddy, what is climate change?"**

Stumped! I guess I expected some tricky subjects somewhere down the line (I'd already tried, and failed, to describe why the sky is blue and how a rainbow works), but seriously? Climate change? At three? Why isn't there a parent's manual for this stuff?!

I am a professional broadcaster and though not by any means the world's finest scientist, I do have an appropriate amount of letters after my name. Trying to convey the natural world's triumphs and challenges in a way that anyone can understand is what I do for a living. Yet climate change has me – and many of my colleagues – stumped.

In many ways, **AMY AND ELLA'S** first passion-project of the single-use plastic crisis is perfect for science communicators – its simple, cause-and-effect message is clear. There is something here that we can all do to create a better outcome. Climate change is the exact opposite. Its complicated cause-and-effect is jumbled – it's difficult, for example, to tie any hurricane directly to global warming. Also, sometimes it seems just too big and complex for any of us to make a meaningful impact. That leads to the worst of all outcomes – young people who feel they don't have a voice and are without power, rather than charged with the desire to go into battle.

So how do we cut through the noise? Is it possible to distil climate change down to a few clear facts? How do we work out the difference between important fact and distracting fiction? Can we individually make a difference? Is there a way to find hope and positivity in all the doom and gloom?

Thankfully, you don't have to rely on me for the answers to any of these questions. Thankfully, we have two superheroes in the form of Amy and Ella.

I was there on my feet cheering with the glitterati of high society when Amy and Ella took their Pride of Britain award. I've spoken alongside them to many important institutions, including the UK parliament. I was there at COP26, where they rode the crest of a wave of young activists setting the world alight (metaphorically speaking!) with their eloquence, enthusiasm, and ambition. I rather think I shall be spending much of the twilight of my career standing in their shadow and learning my lessons from them. And this is how it should be. This generation has a new way of telling their stories that dinosaurs like me will never truly understand. Fitting then, that the champions and storytellers should rise from the ranks of **THIS** generation, the one we are relying on to effect big changes.

These sisters are a unique voice, and an epic instrument for change and lucid thinking.

## REMEMBER THEIR NAMES.

# INTRODUCTION
## by Amy and Ella

You may have already heard – planet Earth has set us some pretty demanding deadlines, and it's important that we meet them:

BY 2030, GLOBAL $CO_2$ EMISSIONS NEED TO BE HALVED.

BY 2050, GLOBAL $CO_2$ EMISSIONS NEED TO BE AT NET ZERO.

You will have probably heard these statements many times before, and in the years leading up to these deadlines you are likely to hear them many more times. And we make no excuse for using them a few times in this book, too. But deadlines, no matter how important they are or who sets them, will not be met if they are not understood by enough people. Without being understood, they are empty... just a bunch of words that aren't taken seriously.

But us kids are smart, right? And we know that if we don't understand something, we need to learn about it until we do – particularly if it is important – like the potential end of humanity as we know it. And when we've understood it, and we realize there's a need to do something about it, we get on and do just that, with real urgency.

It's pretty simple logic to us kids, but for some reason, many adults and leaders around the world don't seem to see it this way. Some people seem to bury their heads in the sand and carry on regardless, while others refuse to accept the reality because they may be clouded by financial or business influences. For some, the truth is just not convenient.

However, we think, and hope, that many people do really care – they just need help to understand big issues like the climate crisis, and to be supported by practical ideas for action and lifestyle change.

That's where we come in. We are the generation that must insist that adults take the urgent action needed to change legislation, influence business, and ensure that we don't miss the deadlines planet Earth has set us. Globally, we cannot afford to put off urgent action any longer, and we're quickly running out of time to make the changes we need to see.

We need to see change happen on the individual and national level. We need to **cut our carbon emissions**, **reduce our consumption**, and **hold the governments and businesses that answer to consumers accountable**. Ultimately, we all need to be a little more **CLIMATE CLEVER**, and hopefully this book will help you to be just that.

We're hoping it will help you to understand the science behind the complicated issue of climate change; to see the changes that you can make to your life, and encourage others to do so as well; and to raise your voice to take wider action against this threat. Hopefully, it will show you the power that you hold to be the positive change that you want, and need, to see in the world.

# 1

## What is
# CLIMATE CHANGE?

OK, let's start with the basics – so what exactly is climate change? We often see it as a complicated and overwhelming issue, stuffed full of technical language that only the smartest people on Earth can understand...

But it doesn't have to be that way! The climate crisis is something that we can – and should – all learn more about, so that we can see the ways we can combat it. So in Chapter 1, we're going to take a look at the basic science, evidence, and causes of climate change.

# Jargon BUSTING

There are a lot of confusing terms used when talking about climate change, and we're going to be using quite a few of them in this book. So, to get started, let's wrap our heads around the key climate lingo.

## Carbon dioxide

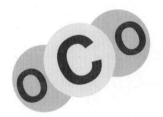

Carbon dioxide ($CO_2$) is a gas found on Earth. One molecule of $CO_2$ contains one carbon atom and two oxygen atoms. Concentrations of $CO_2$ are increasing rapidly in the Earth's atmosphere and this is one of the main causes of climate change.

## Greenhouse gases

Greenhouse gases, including $CO_2$, are the important gases generating climate change. When they build up in the atmosphere, they create something called the **"greenhouse effect"**. What happens is that the gases trap energy from the Sun, causing the Earth to heat up – a bit like a greenhouse, if you think about it!

Some of the key greenhouse gases causing climate change are carbon dioxide, methane, and nitrous oxide.

**Find out more about the greenhouse effect on pages 32–35.**

## Climate change

Climate change is a change in the average conditions of a region –
such as temperature or rainfall – over a long period of time. Climate is
different to the weather we see forecast on TV. While climate change
does affect the weather, the actual change in climate happens over a
period of many years – **it's not the same as having a rainy
day and then a sunny one!**

The Earth's climate has changed before throughout history, but not at
the same rate that we're experiencing currently.

## Global warming

The phrase **"global warming"** is sometimes used to describe climate change, because both terms refer to the average temperature of the Earth heating up. However, climate change doesn't just mean that we're going to get more hot weather – it can also cause things like intense winters or snowstorms.

## Emissions

Humans produce a lot of greenhouse gases from things like cars and power plants. These emissions are building up in the atmosphere and contributing to climate change.

## Carbon neutral and net zero

**"Carbon neutral"** and **"net zero"** are terms that are often used to describe the same thing. However, there is a key difference between them:

**Carbon neutral** is when no extra carbon dioxide is added to the atmosphere. This might happen by reducing the amount produced in the first place, or through carbon offsetting.

**Net zero** is a bit different – while it includes carbon dioxide, it also refers to all the other greenhouse gases that cause climate change. The balance of these gases in the atmosphere has to be kept the same in order to achieve net zero.

## Carbon offsetting

Carbon offsetting means that for every amount of carbon dioxide you release into the atmosphere, the same amount is removed or prevented from being released. For example, some companies pay for trees to be planted to capture the same amount of carbon dioxide from the atmosphere that their business has produced (see also page 110).

# THE HISTORY
## of climate change

Climate change isn't something that happens suddenly, and it's not something that we've just discovered. Let's check out some of the important dates throughout history that have defined the climate crisis that we're experiencing.

### 1712

The first widely used steam train is invented by Thomas Newcomen, a British ironmonger. This fuels the Industrial Revolution, a period in which goods are manufactured in large amounts in factories, instead of by hand as they had been before. The manufacturing is largely powered by coal, and soon spreads outside the UK to the rest of the world.

### 1824

French physicist Joseph Fourier discovers what is now called the **"greenhouse effect"**. He calculates that the Earth is warmer than it should be because of how far it is from the Sun, and that the planet's atmosphere must slow down how fast heat escapes to space.

### 1861

Irish scientist John Tyndall discovers that water vapour and other gases cause a greenhouse effect.

## 1896

Svante Arrhenius, a Swedish chemist, recognizes that increases in the amount of $CO_2$ in the atmosphere from burning coal is causing the Earth's temperature to rise. At the time, people think this is a good thing – it is believed that global warming makes food production easier, and could save us from future Ice Ages.

## 1938

British engineer Guy Callendar publishes research showing that in the last 45 years, the Earth's temperature has risen by 0.5°C. He finds that alongside this temperature rise, the amount of $CO_2$ in the atmosphere has also increased, and suggests that this might be what has caused the warming.

## 1958

US scientist Charles David Keeling runs a four-year project, which measures $CO_2$ concentration in the atmosphere, and provides the first solid proof that this amount is rising.

## 1972

The first major UN environment conference takes place, although climate change is not one of the main issues discussed.

### 1975
US scientist William Broecker popularizes the term **"global warming"** by including it in the title of a scientific paper.

### 1988
The Intergovernmental Panel on Climate Change (IPCC) is formed to research climate change.

### 1998
The Hockey Stick climate graph is published. It shows that the current temperature increase is dramatically different to anything seen in the last 1,000 years.

### 2007
The IPCC publishes their fourth report, which finds that it is more than 90% likely that humans' greenhouse-gas emissions are causing present-day climate change.

### 2011
Data shows concentrations of greenhouse gases are rising faster than in previous years.

### 2015
The Paris Climate Agreement is signed. This is the first legally binding treaty on climate change.

### 2021
Global $CO_2$ emissions reach over 36 billion tonnes per year.

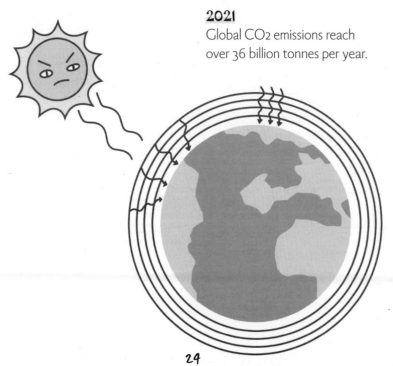

# On this
# DAY

## 11th December 1997

### THE KYOTO PROTOCOL WAS ADOPTED

The Kyoto Protocol was a landmark in the global work to tackle climate change. It was the first-ever legally binding climate treaty, which means that the countries that agreed to it had to, by law, reduce their greenhouse gas emissions. While the treaty was far from perfect and pretty unambitious compared to newer climate agreements, it was a milestone for its time. It was let down by the fact that China and the USA (the two biggest emitters of greenhouse gases) did not take part in it, but the Kyoto Protocol still paved the way for future, more radical, climate agreements.

# What is the
# CLIMATE CRISIS?

Climate change can be a **biiig** issue to understand, so let's start with the basics – **WHAT IS IT?**

Climate change is a change in weather conditions that we experience over a long period of time. It's caused by an increase in the amount of greenhouse gases in the atmosphere, which are released from human activities like driving cars and burning fossil fuels.

There are lots of different greenhouse gases, but the main ones that are contributing to climate change are:

## Carbon dioxide ($CO_2$)

Carbon dioxide is responsible for most of the global warming we're experiencing because humans are releasing lots of it into the atmosphere! And once in the atmosphere, $CO_2$ stays there for hundreds of years.

## Methane ($CH_4$)

Methane is responsible for around 30% of global warming. It is much more potent than $CO_2$ – it absorbs more heat, which warms up the planet at a faster rate. However, methane only lasts in the atmosphere for around 12 years.

## Nitrous oxide (N₂O)

Nitrous oxide has 300 times the impact on warming the planet compared to $CO_2$, but there is a smaller concentration of it in the atmosphere. The biggest source of $N_2O$ is agriculture, in particular animal waste and fertilizers.

## Fluorinated gases

Humans are the only source of fluorinated gases, which are the most potent and long-lasting of all greenhouse gases. Even a small amount of fluorinated gas can have a big impact on the climate!

The increasing amount of these greenhouse gases in the atmosphere is causing the global average temperature to rise. This is calculated by combining the temperature of lots of different places on Earth, and it has already risen by over 1°C in the last couple of hundred years!

The Earth's climate has changed before throughout history — but over a period of hundreds of thousands of years, not just a hundred or two!

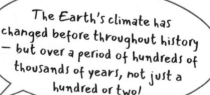

The fact that the Earth has warmed so quickly shows scientists that present day climate change is unnatural and caused by humans.

# CO₂ emissions

**Over 36 billion tonnes of** CO₂ are released per year – that sounds like a massive number, and one that's really hard to imagine.

When talking about carbon dioxide emissions, we often use tonnes as a measurement – but what does 1 tonne actually look like? Well, to visualise this, **let's imagine what 1 tonne of CO₂ emissions is equal to in everyday terms.**

1 tonne of carbon dioxide would look like 500 CO₂ fire extinguishers.

It would be the amount of CO₂ produced per passenger in a return flight from Paris to New York.

And, it would be the equivalent of the emissions from charging 121,643 smartphones.

Now, you might be thinking, **CO2 is a gas? How can a gas weigh anything?**

Well, you're right — $CO_2$ is a gas, but a gas is made up of lots of tiny atoms. All these atoms have a mass, so they all have a weight on Earth. But if that's a bit hard to imagine, picture a bubble of $CO_2$ that is **10 m wide** and **10 m high**. If you put that bubble on a big set of scales and weighed it, that bubble would weigh about 1 tonne.

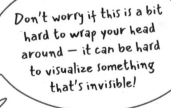

Don't worry if this is a bit hard to wrap your head around — it can be hard to visualize something that's invisible!

Comparing 1 tonne of $CO_2$ to everyday things can help us to understand just how much we're releasing into the atmosphere.

# The CARBON cycle

When talking about the climate crisis, carbon dioxide is often presented in a negative way. After all, it is one of the leading gases causing the greenhouse effect! But, believe it or not, carbon dioxide is not all bad. In fact, it's a big part of an important natural process called the carbon cycle.

The carbon cycle is pretty amazing. It explains the way carbon is stored and moved around on Earth. It contains **SOURCES** – parts of the cycle that add carbon to the atmosphere – and **SINKS** – parts of the cycle that remove carbon.

## Respiration

Animals and plants both respire to produce carbon dioxide ($CO_2$). Respiration is a process where oxygen and sugar are converted into energy, which is used by living things for growth and movement. $CO_2$ is created as a result of respiration. **We breathe out this gas.**

## Photosynthesis

Plants go through a process called photosynthesis to make food. To do this, they need $CO_2$ from the air. This is why plants, such as trees, are really important in fighting climate change because they can absorb lots of $CO_2$ from the atmosphere.

## Decay

When plants and animals die, they are broken down by tiny microorganisms called decomposers. These decomposers also respire, releasing $CO_2$ into the atmosphere.

## Fossilization

If dead organisms don't decompose, they can build up and be compressed over time, and some form fossil fuels. These fossil fuels contain lots of trapped carbon, which is released as $CO_2$ into the atmosphere when humans burn them.

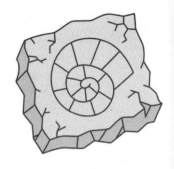

## Volcanoes

Volcanoes are also a part of the carbon cycle – they release $CO_2$ when they erupt. However, at the moment, the amount of $CO_2$ volcanoes release has a very small effect on the global climate.

## The oceans

Oceans are one of the most important carbon sinks. **They take in $CO_2$ from the air!** Find out more about the oceans' role in the climate crisis on pages 36 and 80.

Normally, when us humans aren't interfering, this cycle is balanced. However, we're releasing much more $CO_2$ into the air (through processes such as burning fossil fuels) than can be absorbed. This causes the balance to tip and too much $CO_2$ to build up in the atmosphere.

# The
# GREENHOUSE
## effect

Greenhouse gases are getting a lot of bad press at the moment, mainly because of the huge amount that we're releasing into the Earth's atmosphere. This is causing our planet to warm up very quickly. But, these **greenhouse gases** are actually really important. In fact, we couldn't live without them.

Surrounding the Earth is our **atmosphere** — a layer of gases that shields the Earth from harmful radiation from the Sun. The atmosphere helps to keep the Earth at a temperature that is liveable. It's mainly made up of the gases nitrogen and oxygen, along with a small amount of argon gas. Other gases, such as greenhouse gases, make up just **0.03%** of the atmosphere. Despite greenhouse gases making up such a tiny proportion, they play a big role in trapping heat from the Sun.

This is a good thing IF there is the right amount of greenhouse gases in the atmosphere. Without these gases, the Earth's temperature would be an average −18°C. That's a bit too chilly for me — brrr!

However, if the greenhouse gases build up in the atmosphere to a level that's too high, too much heat is trapped and the Earth warms up too much. This is what's causing the current climate crisis.

## SO HOW DOES THE GREENHOUSE EFFECT WORK?

Sunlight enters the atmosphere. The energy from it is absorbed by the Earth's surface. The Earth then radiates this energy back into the atmosphere.

Some of this energy passes back into space, but some of it isn't able to leave because it is stopped by greenhouse gases. This causes the atmosphere to heat up.

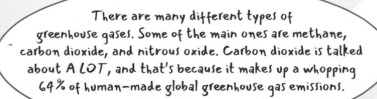

There are many different types of greenhouse gases. Some of the main ones are methane, carbon dioxide, and nitrous oxide. Carbon dioxide is talked about A LOT, and that's because it makes up a whopping 64% of human-made global greenhouse gas emissions.

But it's not just the levels of greenhouse gases in the atmosphere that are important. Different greenhouse gases, such as methane and nitrous oxide, absorb more heat than others. This is why even small changes in the levels of these gases can have a big impact on the global climate.

To complicate things a little, **water vapour** is actually a greenhouse gas, and it's the most common one at that! However, water vapour doesn't control the Earth's temperature, but is instead controlled by it. When the atmosphere is warm there is more water vapour. This is because warmer temperatures make more water evaporate from the Earth's surface into the atmosphere. Since this water vapour acts as a greenhouse gas, it then allows more heat to be trapped – causing even more warming. So, water vapour increases the effects of warming already caused by gases such as $CO_2$. This is called a positive feedback process.

# OCEANS

Water is brilliant at storing heat. It has what's called a **high heat capacity** – basically, it can absorb a lot of heat energy before it starts to get warm.

Oceans cover roughly **70%** of the Earth's surface, so perhaps unsurprisingly they play a large role in controlling the planet's temperature. Without the oceans, the Earth would be much hotter than is it now, because the water spreads heat around the planet. The oceans are also very good at absorbing extra heat from the atmosphere as the planet warms up. In fact, the top few metres of the oceans contain as much heat as is in all of the Earth's atmosphere!

Not only do the oceans play a major role in absorbing heat, but they also help with rising $CO_2$ levels. About **a quarter of the $CO_2$** humans produce is absorbed by the oceans. Plants in the ocean use $CO_2$ to **photosynthesize** and produce oxygen, just like plants on land.

Take a deep breath — believe it or not, half of all the oxygen you just breathed in was produced by tiny ocean plants.

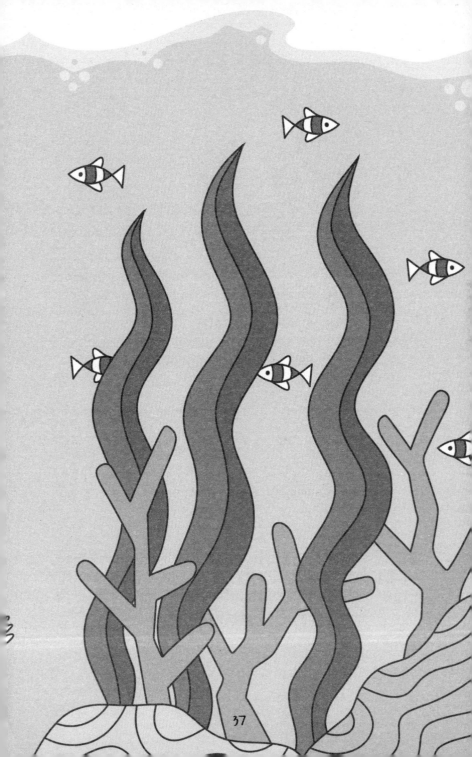

But this all comes at a cost. As well as helping to prevent climate change, the oceans are greatly affected by it. The large amount of $CO_2$ they absorb makes their waters more **acidic**, as the $CO_2$ reacts with water to form an acid. This is dangerous for sea life, as the acidic water can break down the shells of sea animals. It can also damage and bleach coral reefs. **Coral bleaching** is what happens when the algae that coral feeds on dies, causing the corals to be weakened. This is bad news, as coral reefs are home to lots of living creatures.

Plus, as the oceans warm up from the heat of the climate crisis, ocean currents moving cold and warm water around the planet could slow down and stop. While this might not sound like a big deal, it could change the regional climate in places like Europe, which is normally mild due to warm ocean currents.

To top it all off, as ocean water heats up, it expands. This means it takes up more space in the oceans, causing **sea levels** to rise. Melting land glaciers and ice sheets flow into the ocean, which also increase the water level.

SO, IT'S NOT LOOKING GREAT FOR OUR OCEANS - DESPITE HOW IMPORTANT THEY ARE TO OUR SURVIVAL AND IN SLOWING DOWN THE EFFECTS OF THE CLIMATE CRISIS.

# FAKE NEWS

## CLIMATE CHANGE CAN'T BE BAD FOR THE ENVIRONMENT: PLANTS NEED $CO_2$ TO LIVE!

Yep, it's true – plants do rely on $CO_2$ for processes that they need to survive, and therefore are responsible for removing and storing large amounts of carbon dioxide from the atmosphere. The problem is, because humans are releasing such a huge amount of $CO_2$ – much more than would naturally be produced – the plants and trees on Earth can no longer keep up and remove it at the same rate it's entering the atmosphere. So, don't worry, reducing carbon emissions won't kill all the plants on Earth – they need nowhere near the amount that's currently being produced!

# CAUSES
## of climate change

So, we've caught up with some of the science of climate change – how greenhouse gases, like $CO_2$, are causing the planet to heat up.

But where exactly do these greenhouse gases come from?

Well, some of them are produced naturally in the **carbon cycle**, such as carbon dioxide from **respiration** and **decomposition** (page 30). Some greenhouse gases even come from **volcanoes** – when they erupt, they release carbon dioxide and other gases into the atmosphere.

But these natural sources of greenhouse gases are dwarfed in comparison to emissions from **human activity**. People are responsible for producing **100 times the CO2 of volcanoes** (mainly, by the burning of fossil fuels). These activities happen for a reason, to provide us with the products, the transport, and the food we need to live our lives. And while they've made our lives easier and more comfortable, they come at a big cost — messing with the global climate.

There are over 7 billion people on the planet at the moment, and they all need certain things to live — like food, shelter, and fuel.

But the way we live our lives in wealthier countries means that we're using too many resources. They all need to be made somewhere and somehow, and that's taking a big toll on our planet.

Let's take a look at the sectors that are releasing the majority of the greenhouse gases warming up our planet.

# ENERGY

Fossil fuels are what's called **"energy dense"**. This means that burning even a small amount of a fossil fuel can release a whole lot of energy. Because of this, they're a valuable resource for generating energy that we rely on to go about our lives. We use fossil fuels to heat our homes, to produce electricity, and so much more. In 2019, around 84% of global energy came from fossil fuels.

The problem is that using large amounts of fossil fuels to generate energy also releases large amounts of $CO_2$ into the atmosphere. Fossil fuels are formed from the ancient remains of organisms, and contain lots of trapped carbon. When fossil fuels are burned, this carbon is released into the atmosphere as $CO_2$.

## THERE ARE THREE MAIN TYPES OF FOSSIL FUEL: COAL, OIL, AND NATURAL GAS.

Coal is mainly used to produce electricity. It is burned in power stations to heat water and produce steam. The steam then makes a turbine inside the station spin, which turns a generator, making the electricity we use. This creates electricity, but also lots of $CO_2$. Coal alone is responsible for almost a third of the global temperature increase that's happened so far.

Oil is used for a range of purposes. It can be converted into different petroleum products (a process called refining), which are used to generate electricity, heat buildings, fuel cars, and create items like plastics. Oil provides 40% of the world's energy, but this comes at a cost. Burning and refining oil releases **greenhouse gases** into the atmosphere, along with other types of air pollution. The process of drilling for oil can be dangerous, too, with oil sometimes being spilled into the oceans where it can kill lots of wildlife.

Natural gas is mainly made up of methane, which burns more efficiently than coal, meaning that it releases less $CO_2$ and air pollution. However, you might remember that methane is also a greenhouse gas – so, when some of this methane leaks into the atmosphere during the production and transportation of natural gas, it remains in the atmosphere for a long time, adding to the greenhouse effect.

To tackle the climate crisis, we need to reduce the amount of fossil fuels that we use, and produce energy in eco-friendly ways. Find out more on page 96.

# TRANSPORT

In today's world, people travel a lot – whether that's away on holiday, visiting family or friends, or travelling locally for shopping or work. Because of this, we have many different ways to get around, though not all are so good for the environment. Transport now accounts for around **a quarter of Europe's yearly $CO_2$ emissions**. In countries like the UK and USA, it's this sector that produces the most $CO_2$.

Airplanes get a lot of bad press due to their environmental impact. **Emissions from flights only account for around 2% of human $CO_2$ emissions**, but this is still of concern because flights are one of the fastest growing sources of $CO_2$. Also, flying isn't accessible to everyone – the 1% of the world's population that fly lots cause over half of the yearly flight emissions. Plus, just one flight can release a lot of $CO_2$. For example, the emissions from a single flight from London to New York are more than the average amount of $CO_2$ released by a person in Paraguay in an entire year!

So, to prevent emissions from transport, we stop flying, right? Unfortunately, it's not that simple. Flying has many advantages over other forms of transport. Flying is quicker and cheaper, especially over long distances.

And flights are used for more than just carrying people. They transport goods around the world that need to be taken from place to place quickly, such as fruits and vegetables.

But, before we blame all the transport emissions on flights, it's worth knowing that other transport methods aren't great either. **Shipping products around the world releases about 2.2% of global $CO_2$ emissions**, which is about the same as the whole of Germany does!

How eco-friendly transportation is often comes down to how many people each vehicle carries. A study by the European Environment Agency found that **cars can be more polluting per person than airplanes** if a car travelling a long distance has only one person in it. This is why other methods, such as trains and buses, have such a small amount of emissions compared to other vehicles. These modes of transport normally carry a lot of people, so the emissions per person are much smaller.

# AGRICULTURE

Believe it or not, what we eat has a big effect on the climate. Agriculture – **the process of growing crops and raising farm animals** – is responsible for a large amount of emissions, particularly of the greenhouse gases methane and nitrous oxide.

Rearing livestock – that is, raising farm animals such as pigs, sheep, and cows – **makes up 14.5% of global greenhouse gas emissions**. And, cows alone are responsible for two-thirds of this, due to the amount of the greenhouse gas methane that they produce.

So what is it about cows that make them produce such a lot of methane? Believe it or not, it comes down to their burps!

Now, only one cow burping and farting its way through life would have pretty much no impact on the climate. But the problem is, there's more than 1.5 billion of them on Earth. And together, they produce a lot of methane.

The greenhouse gas emissions from agriculture go further than just cow burps. All animals raised for humans to eat need a place to live, and it's often rainforests that pay the price. Since the 1960s, an area the size of Sweden has been cleared for raising cows in the Amazon rainforest – a habitat that produces **20% of the world's oxygen**.

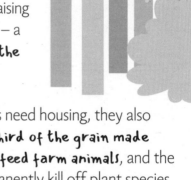

And not only do these animals need housing, they also need feeding. **Around one-third of the grain made around the world is used to feed farm animals**, and the **overgrazing** of land can permanently kill off plant species that previously grew there.

Before we blame all agricultural emissions on animals alone, it's worth pointing out that growing crops to feed humans can also be damaging to the planet if it is not done sustainably. Artificial fertilizers, which are put on crops to help them grow, can wash off fields into rivers and lakes, damaging the ecosystems and creating **"dead zones"**, where nothing can survive. And, certain types of crops, **such as rice**, can produce emissions themselves. **Rice is responsible for 1.3% of global greenhouse gas emissions.** Making space to grow all these crops also leads to deforestation for monoculture crops – but more on that on the next page.

# DEFORESTATION

Forests cover about **30%** of the land on Earth and are one of the most valuable resources the planet has. They provide a **home to an estimated 80% of the world's land animals and plants**, and are relied on by over 2 billion people for shelter and livelihoods. Forests also play a vital role in combating climate change. A single tree can absorb as much as 21 kg of $CO_2$ from the atmosphere each year. Yet despite how important forests are, huge areas of trees are being cleared every year to make room for less biodiverse ecosystems – **this is called deforestation.**

One of the key culprits of deforestation is **agriculture**. While forests are cleared to make way for the grazing of cattle for meat and dairy foods, nearly half of global deforestation is to clear space for crops, in particular soy. Soybeans are used for a range of human products – from tofu to oils – but the large majority of them (80–90%) are used to feed farm animals, particularly poultry and pigs.

On top of soy crop growth, unsustainable palm oil production has driven the deforestation of some of the most biodiverse forests in the world, mainly those in Indonesia and Malaysia. **Palm oil is used in just about everything** – in fact, 50% of packaged products in supermarkets contain it. The problem is, lots of this palm oil is grown using monoculture crops, meaning only one type of crop is grown in an area. This decreases biodiversity and the habitat for different species.

Logging operations that provide **timber** are also a key driver of deforestation. Huge areas of forest are often cut down for just a few high-value trees, and roads need to be built to access more and more remote forests – which leads to further deforestation.

The World Resources Institute says that if tropical deforestation was a country, it would be the third highest producer of $CO_2$ emissions, behind China and the USA!

When trees are cut down, they release some of the carbon stored in them as $CO_2$ into the air. Cutting down trees also means that they can no longer absorb $CO_2$ from the atmosphere.

# INDUSTRY

Industry is, unsurprisingly, a large cause of climate change. After all, it was the **Industrial Revolution** in the 1800s that first drove a widespread use of fossil fuels, when fuels like coal were burned at a large scale in the move towards mass-producing products. Since then, the use of fossil fuels in producing the items we buy and creating the buildings we live in hasn't slowed down – far from it.

**Industry** basically describes the processes and businesses that make the products and materials we use every day.

There are two types of industry emissions: those that are produced at the factories where these materials are produced (known as **direct emissions**), and those that happen somewhere else but are still linked to the factory (known as **indirect emissions**).

Most direct emissions are produced from burning fuels for power and heat in factories. For example, producing materials like steel, iron, and cement releases lots of $CO_2$ emissions because loads of energy is needed to heat up the materials, meaning fossil fuels are burned.

A smaller number of direct emissions also come from leaks when extracting fossil fuels, or from chemical reactions, such as in the making of cement. One main material in cement is lime (no, not the fruit!), which is converted from another material chemically. During this process, $CO_2$ is produced as an unwanted product.

Indirect emissions are those that are released from burning fossil fuels at power plants to make electricity, which is then used in factories and machines. While these emissions aren't produced directly at the factories themselves, they're still related to industry.

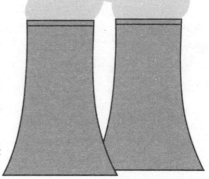

With the world's population still growing and rapid economic growth in some countries, the demand for products and housing is unlikely to slow. That's why it's important for companies to make their processes more sustainable, through new technology and innovations.

# SURPRISING causes

Fast fashion, plastic pollution, air pollution — they're all completely different problems. Surely they can't be related to one another, or to the climate crisis, right?

Wrong! Climate change is closely linked to all sorts of environmental problems, even though on the surface they might seem totally unrelated. Let's take a look at some examples.

## Plastic pollution

With a truck's worth of plastic entering the world's oceans every minute, plastic is causing damage to the planet, killing wildlife, and piling up on beaches. But aside from litter, plastic is closely linked to the climate crisis in two main ways:

Firstly, in its production. Plastic is mainly made from oil, a fossil fuel. The process of taking oil from the ground and refining it releases greenhouse gases (mainly $CO_2$). In fact, oil refinery is the second highest-ranked sector in terms of emissions per facility – behind only the power plant! As the amount of plastic we use grows, so will the oil needed to make it. By 2050, plastics alone are set to be responsible for nearly half of the world's total oil demand!

The other way that plastic impacts on the climate is through litter. Plastics break down over time into tiny **microplastics**, pieces of plastic so small that they can be eaten by microscopic plankton. Plankton play a key role in what's called the ocean "carbon cycle". Normally, the waste from plankton sinks in the ocean, and any carbon this waste contains is trapped in the deep sea. But, plastic makes this excretion more buoyant, which means that it sinks down more slowly to the seabed, or doesn't sink at all. So there's more opportunity for the carbon in these remains to be released back into the atmosphere instead of being trapped on the ocean floor!

## Fast fashion

Fast fashion is the term used to describe clothes that are made cheaply and quickly, and are only worn for a short time before being thrown away, which is bad for people and the planet. Often, people making clothes work for long hours in dangerous conditions, being paid as little as £2 per hour. And when it comes to the environment, the fashion industry releases more emissions than shipping and flying combined, being responsible for a whopping 5% of global greenhouse gas emissions!

But crazily, this huge carbon footprint isn't mainly due to the transport of clothes or even the process of making them, but largely comes down to the very materials that they're made from. Plastic and fast fashion have one key ingredient in common – oil! Over 60% of clothes produced are made up of human-made fibres that come from oil, with polyester **(a type of plastic!)** being most widely used.

And it's not just the human-made materials that are to blame. Growing cotton to make clothes requires water and often chemical fertilizers. Many fertilizers contain carbon, meaning that when they are used they produce lots of $CO_2$!

# WHO are the

# climate change sceptics?

Despite the strong evidence of climate change being caused by humans, there are still some people who question whether it really does exist, and if it's as bad as the science suggests. Let's take a look at some of the main climate change sceptics and the tactics that they use.

## Fossil fuel industry

The fossil fuel industry has a lot to gain from delaying action against climate change – after all, it's the materials they sell that are the key driver of the climate crisis! Crazily, fossil fuel companies are aware of the impact fossil fuels have had on the climate since at least the 1970s, and yet have spent the majority of the years since then spreading fake information about climate change! They either denied it altogether, or pushed the responsibility to take action onto individuals. In fact, it was an oil company that first promoted the idea of a carbon footprint back in 2004, to get individuals to reduce their impact on climate change – all while fossil fuel companies were still drilling for oil and polluting the atmosphere.

# Conspiracy theorists

Some people don't trust scientists – at all. Conspiracy theorists believe that climate change has been made up by scientists and the government, either to get lots of money to do research or to control the public.

Even the 45th US president – Donald Trump – seemed to buy into this theory. He claimed that climate change was a "hoax" (it's not) which was invented by China (it wasn't), and that the noise from wind turbines can cause cancer (there's no evidence to back this up).

## The media

The majority of TV, newspapers, and books around the world don't deny that climate change is happening or is caused by people. So how is it that a study in 2020 found that statements given to the press from those opposing climate action get more coverage than those for it? Well, while some media companies are still owned by people who have a vested interest in denying climate change, that's not the main reason. Rather, it's because of **both-sidesism** – media platforms trying to avoid bias by presenting believing or not believing in climate change as equal arguments. In reality, both-sidesism is misleading, with well over 90% of climate scientists agreeing that climate change caused by humans is real.

55

# WHY deny climate change?

It can be hard to understand why people want to deny climate change. Especially as kids, we see how the crisis will affect our future, and it can be frustrating when people still deny or downplay its impacts. But there are a variety of reasons for this.

## They don't properly understand it

We get it – climate change is confusing. For some people, **climate change denial** might come from not fully understanding what the climate crisis is, or what its effects are. An example of this is where sceptics use heavy snowfall or cold temperatures as evidence that "global warming" doesn't exist, when in reality both of these things can be made more extreme by the climate crisis. Similarly, some climate deniers argue that climate change is actually completely natural, and that scientists are putting too much emphasis on $CO_2$ as the cause.

## They're scared

Eco-anxiety (page 118) is an understandable feeling when hearing about climate change. The difference is, some people react to it by turning that fear into denial of the facts. It's a bit of a bury-your-head-in-the-sand approach. If you ignore what's happening, it might make you feel better in the short term but it doesn't stop climate change from getting any worse!

## They have a vested interest

£ $
€

Vested interest means you have a reason to do something for your personal gain. Often, this vested interest is money – fossil fuel companies pay some politicians, scientists, or other professional people to spread confusion around climate change to get people to question whether it exists. These companies themselves have a vested interest in preventing action against climate change, since the fossil fuels they sell will not be used if the world tries to reduce carbon emissions – meaning they'll lose a lot of cash!

> And it doesn't have to be about money...
> Business people might want to keep their organization happy, or they might be good friends with another high-ranking professional, or a country's leader who's responsible for lots of emissions.

## It doesn't affect them personally

The effects of climate change are already being felt around the world, but not in equal measure. People in some countries experience terrible effects of the climate crisis more often than others, such as through wildfires or flooding. If you haven't had a climate-change-related fire rip through your back garden, you might be less likely to believe climate change exists than someone who has lost their house or family to it. But this doesn't mean that climate change isn't real. The chances are, we'll all feel the impacts of the climate crisis in the future if enough isn't done to stop it.

57

# ARGUMENTS
## that sceptics use

How do climate change sceptics manage to convince people that climate change isn't real? Here are some techniques to be aware of that sceptics use to seem reputable when talking about climate science:

### Using scientists of different specialisms

Someone may be an expert in their field, but not be a reliable source on climate change — that is, not as reliable as a climate scientist. However well-qualified, their subject might not be related to the climate crisis.

# Cherry-picking data

Cherry-picking in science is where you focus only on data you like, and ignore any that you don't. It's like only looking at one small piece of a jigsaw, instead of the whole picture. The vast majority of scientific data suggests that climate change is caused by humans – but that doesn't mean it can't be cherry-picked to suit the argument climate sceptics want to make!

**The planet is complicated** – the global climate isn't just impacted by greenhouse gases, but also by particles in the air called aerosols (natural or human-made) and climate patterns. Because of these other factors, the temperatures in some years can be unusually high or low. For example, in 1998 an El Niño – a natural occurrence that causes some ocean waters to be warmer than usual – made the average temperature spike even higher than expected. Because of this unusually warm year, the increasing average temperature of the years that followed wasn't as obvious, which meant some climate deniers used it as evidence that global warming had stopped. Yet this wasn't the case at all – it just looked that way because the temperature it was compared to was a record high.

**This is why climate scientists always look at temperature change over a long period of time, not just a few years.**

# Doomism

If you think that it's already too late to save the planet from climate change, you're not likely to do anything to tackle it. But giving up plays right into the hands of the fossil fuel companies who continue to pollute the planet. **While the climate crisis is scary, and there is limited time left, it is definitely not too late – the most important thing is to carry on working to stop it.**

# Are COMPANIES to blame?

So where have these emissions come from? We know the sectors that are releasing the most greenhouse gases, but I haven't personally been out digging for oil in my garden! So, who has?

Let's look at who's been responsible for releasing the largest number of emissions throughout history.

In 2019, a report found that just 20 companies were responsible for over a third of all the greenhouse gas emissions produced in the energy sector since 1965!

## AND SURPRISE, SURPRISE – ALL OF THEM ARE COMPANIES THAT PRODUCE FOSSIL FUELS!

Some of them are privately owned corporations. However, a large amount are "state-owned". This means that the companies are either mostly or fully controlled by a government. However, it gets interesting when you look at where these greenhouse emissions came from in particular. Most (a whopping 90%) are produced when the fuels they make are used – such as when we use their petrol to power our cars, or when their coal is used in power plants. Only a small amount comes from the making of the fuels themselves. Now, this doesn't mean that these companies aren't to blame for their emissions – remember, they've been producing their fuels for decades since they found out the damage they were doing to the planet. But since we as individuals are involved in their emissions, we can't just put our feet up and hope these companies will do something.

We can vote with our wallets, and support energy companies that produce electricity from renewable sources (pages 96–105) to play a part in putting these polluting companies out of business. But even more importantly, we can work to push governments to act and make sure that the pollution these companies release comes at a cost for them, and that governments stop their own harmful practices (pages 112–117).

SAY NO TO FOSSIL FUELS!

No more fossil fuels!

# COUNTRIES and
## climate change

**USA:** The USA is the world's second-largest $CO_2$ emitter, with most of it coming from transport and energy production. It's also responsible for the most $CO_2$ emissions throughout history.

**THE GAMBIA:** This country contributes only a tiny amount to the climate crisis and has unique plans to limit its emissions through how it farms. Yet, The Gambia is greatly impacted by climate change. Sea-level rise means that salt water now flows up the country's major river in the opposite direction, so the country suffers from a lack of fresh water.

**INDIA:** This country ranks third-highest in terms of $CO_2$ emissions, largely because of coal use. But the emissions per person in India are low. India is also greatly affected by the climate crisis. Many people have been killed by extreme heatwaves in recent years.

**EUROPE:** Most countries in Europe have taken steps to reduce $CO_2$ emissions, with the largest emitter, Germany, having decreased its emissions since 1990. But Europe was responsible for producing lots of $CO_2$ in the past — for example, until 1880, over half of the world's total released $CO_2$ came from just the UK!

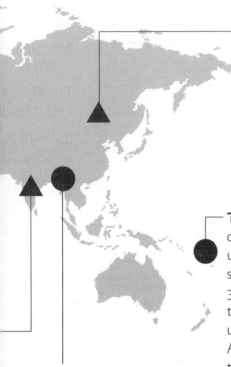

**CHINA:** China has the largest population in the world, and an economy that's growing quickly. Because of this, it is the country that releases the greatest amount of $CO_2$ — over a quarter of global emissions! But, the emissions per person in China are comparatively low.

**TUVALU:** This tiny island-nation could be at risk of being uninhabitable by 2030 due to sea-level rise. The islands rarely go 3 m above sea level, meaning that they're at risk of being submerged under water by the rising ocean. Ahead of the COP26 summit in 2021, the Foreign Minister of Tuvalu delivered his speech up to his knees in the sea to show the effects of climate change on the islands.

**BANGLADESH:** Bangladesh releases only a fraction of the emissions of the world's largest polluters, yet it's impacted much more by the effects of climate change. In 2020, almost a quarter of the country was flooded, meaning millions of people lost their homes.

**KEY**

 = Countries that release the most $CO_2$ emissions

 = Countries that are most affected by $CO_2$ emissions

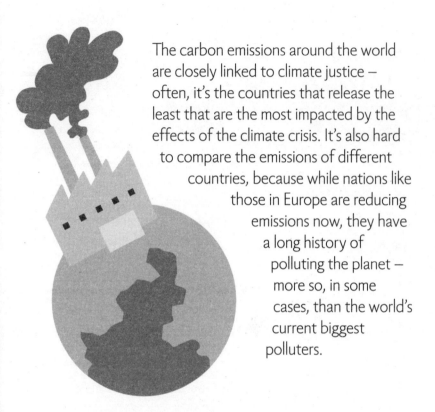

The carbon emissions around the world are closely linked to climate justice – often, it's the countries that release the least that are the most impacted by the effects of the climate crisis. It's also hard to compare the emissions of different countries, because while nations like those in Europe are reducing emissions now, they have a long history of polluting the planet – more so, in some cases, than the world's current biggest polluters.

But the most important thing of all is to remember that this isn't about finger-pointing – it's no good blaming everything on one country or another and saying, "Well, I'm not going to change until they do!" All that does is delay action against climate change, when we need to work together to reduce emissions NOW.

# What is the EVIDENCE?

While a very small amount of scientists dispute whether climate change has been caused by humans, 97% of climate scientists are in agreement that humans are responsible for the current climate crisis.

But how do we know that the climate now is different to how it was 100, or even 1,000 years ago? There's no one alive on Earth that's old enough to be able to tell us that it was a bit colder back in 1800 than it is now (that is, unless vampires are actually real!).

Luckily, some clever people have found ways to measure the change in temperature, weather conditions, and the amount of $CO_2$ in the atmosphere over time. These can show a link between greenhouse gases and the change in the climate. That means that we don't need to rely on immortal people to tell us how the Earth has changed — the science does that for us!

# TEMPERATURE
## change

One of the main methods scientists use to show that the climate has changed, is to look at the **average global temperature** throughout history.

There's one graph in particular that's used a lot, and it's called the **Hockey Stick graph** for fairly obvious reasons! It was published in 1999 by Michael Mann and two of his colleagues. It shows how global temperature has changed over a long period of time – roughly the last 1,000 years, and how it has suddenly accelerated! **This indicates that the changes in the climate that we're seeing at the moment isn't natural, and that humans have played a big part in it.**

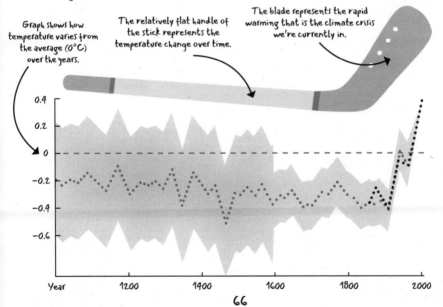

Graph shows how temperature varies from the average (0°C) over the years.

The relatively flat handle of the stick represents the temperature change over time.

The blade represents the rapid warming that is the climate crisis we're currently in.

When the Hockey Stick graph was first released, it got lots of claims from climate deniers and fossil fuel representatives saying that it was incorrect, or used evidence that was not suitable. But, since the graph was released, there have been lots of studies by other scientists that show a very similar result: that as the amount of greenhouse gases in the atmosphere rise, so does the global temperature.

Plus, there have been recent studies that have extended the temperature records of the Hockey Stick graph back to the last Ice Age, and have found that the Earth is warming up faster now than it has in the last 11,000 years!

The data for the Hockey Stick graph was taken from a range of different sources, from tree rings to ice cores to coral! Find out more about how these work on page 70. To calculate the temperature change of more recent years, scientists don't need to backdate the evidence, as they can collect it as it happens.

This is done by measuring the temperature of the air above the surface of the land and the ocean, and calculating how different it is to what is expected. The differences are called anomalies. These can then be used to find out the temperature anomaly on average for a month, season, or even a year!

# The INDUSTRIAL Revolution

You might have noticed that on the Hockey Stick graph, the temperatures start to spike in the 1900s. This shows the effects of the Industrial Revolution — when people began to burn fossil fuels like coal and oil to power machines and transport.

## THE INDUSTRIAL REVOLUTION BEGAN IN THE UK IN THE 1800S . . .

. . . before spreading to other parts of the world. By the end of the 1900s, the majority of the world was reliant on fossil fuels. This increased use of fossil fuels meant that carbon dioxide emissions grew rapidly throughout this time, which scientists think has also caused the warming we've seen in recent decades.

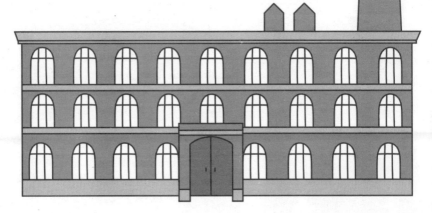

# Gathering
# EVIDENCE

How do scientists get the data needed to show changes in the climate? There are a few different forms of evidence that scientists use to calculate important information as to what the global climate used to be like.

## Tree rings

Have you ever seen a chopped tree stump in a forest and the beautiful circles that cover the inside of it? Those are **tree rings**, and aside from being really cool to look at, they can tell us a lot about what the climate used to be like. Trees are sensitive to the conditions in which they grow, such as temperature and rainfall. If it's a warm and wet period, the tree rings will be wider. If it's cold and dry, they'll be thinner. Or, if there's been drought and other harsh conditions, the tree might not grow at all. So, by looking at the different thicknesses of the tree rings, scientists can work out what conditions were like in the past.

A tree also has a ring for every year it has been alive, so by counting the rings, scientists can work out how old it is! Some of the oldest trees alive are thousands of years old, and so they can tell scientists what conditions were like over the past hundreds of years.

# Ice cores

Glaciers and ice sheets are formed from years and years of snow. Every year that snow falls, it pushes down on the snow below, making it compact enough that over many years it forms ice. Now, prepare to be wowed – when the snow fell on these glaciers throughout history, the temperature of the air left an imprint on the water molecules! And not only does this ice trap the snow that fell, but also particles called **"aerosols"** from the air, and even tiny bubbles of air that can show what the atmosphere contained at the time the air was trapped.

> The oldest ice cores occurred 800,000 years ago – how awesome is that?

Scientists extract ice cores from glaciers to analyze what the climate was like in the past, and use the information to improve models of what the climate might be like in the future.

# Coral

The oldest coral reefs on Earth are millions of years old, and are very sensitive to changes in the climate. That's why **coral reefs** are damaged so greatly by climate change. Corals grow by forming skeletons out of a mineral called calcium carbonate, which they make from ocean water. The density of this mineral – basically, the amount of it packed into a certain area – is affected by things like temperature and light. This means that changes in the coral density produce rings, like those in trees! Scientists can use the rings on coral to work out how old it is, and what the conditions were like when it grew.

71

# 2

## What are the
# IMPACTS?

The world is changing. We're seeing extreme weather, loss of habitats and land, and scorching temperatures, all as a result of the greenhouse gases humans have been pumping into the atmosphere for the last couple of hundred years. And as the planet changes, so do the lives of millions of people and animals around the world.

Luckily, there are some amazing people working hard to combat the impacts of climate change on all sorts of areas – from tackling a lack of water availability, to protecting the animals most threatened by climate change, to researching the climate itself. Since they're the experts, in this chapter, we're letting them explain the impact climate change is having on their field of work, and the change that they want to see to combat it.

73

# NIALL McCANN

With habitats and conditions changing due to climate change, many animals are struggling to survive.

Niall McCann is a biologist and explorer who has presented award-winning documentaries about the environment around the world. As Director of Conservation at National Park Rescue, he has been working to prevent poaching, and is passionate about helping to protect the world's endangered species.

### How is climate change impacting wildlife?

Some of the impacts on wildlife are immediate, such as animals getting caught up in fires or flooding. Other impacts are felt over years and decades. Some animals are actually faring better due to climate change, with warmer temperatures allowing them to live in places that were previously too cold to survive; while others are faring badly, unable to adapt to changes in food availability or temperature.

## What kinds of animals are affected the most?

The most obvious are those that live in colder parts of the Earth. Polar bears rely on the annual formation of sea ice so they can hunt their favourite food: seals. Climate change is making the sea ice form later and break up earlier, giving the bears far less time each year in which to get all the food they need to survive through the barren summers. Any type of animal that can't move is also affected by climate change. When temperatures get really hot, most animals can escape to somewhere cooler, but animals such as mussels and oysters can't run away. The most gruesome example of this in recent times was the "heat dome" of 2021. This affected the Pacific coast of the USA and Canada in the summer, killing over a billion sea creatures, which were literally cooked alive in their shells.

## Will climate change cause all of the Earth's wildlife to go extinct, or can animals adapt?

Animals are amazing at adapting to new conditions —they have to be, because our planet has experienced incredible changes in climate many times. Around 20,000 years ago, most of Europe was living in the Ice Age. Many of the species that roamed the Earth back then, such as woolly mammoths, are now extinct. But many millions of other species have survived the warming of the planet that has taken place since that time. There have been five previous major extinction events in the Earth's history, all of which have been caused by rapid and major changes to the world's climate. The most devastating of these, the Permian-Triassic Extinction Event, caused the extinction of up to 98% of all species on Earth. We are currently living through the sixth major extinction event as a result of climate change, habitat loss, and hunting. This period is likely to cause the extinction of many millions of species, possibly even our own. However, you can guarantee that many millions more will survive, some of which will come to dominate the Earth in the millennia to come!

# VEGA OSMAN BISSCHOP

Vega Osman Bisschop is a youth ambassador for Real Food Systems, through which she and her fellow ambassadors raise awareness of how eco-friendly food systems can help tackle climate change. In school, she and her friends organized a meeting with students and staff members to plan for more sustainable school lunches, and she's gone on to spread her message at an EU panel on the effects of climate change on people around the world. She tells us how climate change and food are linked, and how food systems can be made more sustainable.

## What is meant by a food system?

Food systems are rather like farms. The farmers take care of their crops and animals, and keep the machines running. A food system, like a farm, is made up of different parts working together. Everybody and everything that plays a role in getting a seed in the ground to become a meal on our plates is a part of this food system.

# How is climate change impacting global food systems?

Growing food relies on weather patterns that humans have observed over thousands of years. Knowing when it's going to rain more during the year lets us decide what plants to grow when. Today though, climate change is making these weather patterns more unpredictable, which makes growing our food more difficult in many places. In the oceans, rising temperatures are threatening the fish we catch for food, and on land, frequent storms are damaging crops before they are ready for harvest. Our food sources are becoming harder to rely on.

# Why does climate change damage food systems?

Food systems depend on the Earth's natural environments. Everything we eat comes from the soil or the oceans of our planet, which is now experiencing strong climate impacts. All the natural elements around us, from the air we breathe to the minerals in the ground, and from the water in the rivers to the algae in the oceans, are being affected and are changing our food supplies. Since food systems are made up of many connected parts, when one part is changed, the others are also changed.

# How will food systems need to adapt to climate change?

Since weather patterns are changing, our methods of growing food must also change. Hotter summers in some parts of the world might mean that plants that are more resilient to these temperatures will have to replace old crops. We also need to rethink the fuel-dependent farming methods we use, such as big, fuel-guzzling tractors, and replace them with more climate-friendly alternatives.

# How can food systems be made more sustainable?

Scientists are exploring new ways with food systems. Vertical farms, where fruit and vegetables are grown on walls, are becoming quite popular. Intercropping, already practised by some Indigenous people, involves cultivating certain vegetables next to each other, such as tomatoes and lettuce, which together nourish the ground.

# KATIE ALCOTT

Water is essential to life, and yet one in ten people around the world don't have access to safe water. Climate change is set to make this problem worse, threatening the world's water and making more people unable to access water.

Katie Alcott MBE launched the charity Frank Water in 2005 to help tackle this growing problem. Since then, she has worked to improve hygiene and access to water in India and Nepal.

## What is water insecurity and who is affected?

Water insecurity is where a person does not have access to enough water for all of their needs, all of the year. This affects up to 4 billion people on the planet, many of whom have at least one month of the year where their water supply is insufficient. Those most affected are the poorest in society. Even in poorer countries, rich people have access to water whenever they need it, while the poor struggle to meet their needs. Geographically, there is a "triangle of thirst" running from the Horn of Africa to Pakistan and back to southern Spain, where up to 2 billion people live in a water-insecure environment.

## How does climate change affect global water resources?

Climate cycles and water cycles are in many ways the same thing, and as one changes so does the other. This is seen through the increased number of water-related natural disasters that we are experiencing as the climate is changing. According to UNICEF, "Around 74% of natural disasters between 2001 and 2018 were water-related, including droughts and floods. The frequency and intensity of such events are only expected to increase with climate change."

## How will the changing water cycle impact lives?

As the water cycle changes, the impact falls on the poorest in the world. This is because natural disasters such as floods and mudslides occur in areas where cheaper, unplanned housing is located. People on lower incomes are also the least able to recover from natural disasters due to the lack of resources available to help them rebuild and move on. Research has shown that disasters affect women more than men, so this gives an additional impact on the most marginalized people in the world today – women living in poverty.

## What can be done to combat the effects of climate change on water availability?

We need to improve the way we manage water so that it meets the needs of all people and, crucially, use it sustainably into the foreseeable future. This type of work is part of the Sustainable Development Goals (SDGs) and is set out in detail under SDG 6.5., which focuses on Integrated Water Resources Management (IWRM). At Frank Water, we work to promote IWRM and encourage local communities, NGOs, and governments to focus on understanding the entire water system. As many people have said, climate is water, and so by better understanding how we can use our water responsibly, we are moving in the right direction with our climate, too.

# Interview with...

# EMILY PENN

Emily Penn is an artist and skipper who is dedicated to protecting the oceans. She's organized the largest-ever community-led litter pick on a tiny Tongan island in the Pacific Ocean, travelled around the world on a biofuelled boat, and collected samples of plastic particles in the sea. She also runs eXXpedition, a series of boat trips with all-female crews that research how plastics can relate to female health.

## How is climate change impacting the oceans and the wildlife that lives in them?

The ocean covers more than 70% of the surface of our planet, and ocean currents regulate our global climate. Climate change causes the ocean to become warmer, sea levels to rise, and leads to ocean acidification (which means the water becomes more acidic).

Much like pollution in our ocean, this has huge implications for marine life. For example, sea-turtle nesting beaches and other shallow-water marine nurseries are lost as the sea level rises. Warming seas and acidification force marine life to migrate to new areas, impacting their development, and decimating the food chains on which they rely.

## Why are the oceans damaged so much by the climate crisis?

Our atmosphere and oceans are intrinsically linked. As greenhouse gases trap more energy from the Sun, the oceans absorb more heat, which leads to them becoming warmer and sea levels to rise, which will be devastating for the planet. For us land dwellers, the impact of climate change and pollution on our oceans is often an unseen crisis, because what's out of sight is out of mind – that's why I've been on a mission to make the unseen seen! I want to inspire action.

## Are the changes happening to the oceans permanent?

On my ocean voyages, I've seen just how resilient our ocean is. Coral reefs in the remote Pacific Ocean that were once bleached by a rise in sea temperature are now abundant with life. If we can take away the pressures of climate change, pollution, and overfishing, the oceans will bounce back and recover. But we do need to take action now.

## How can we help to protect the oceans from climate change, and why is it important to do so?

Every bottle and toothbrush polluting the oceans once belonged to someone. It's billions of micro-actions (small, everyday acts) that have led us to this situation, and it's micro-actions that will get us out of it. Everyone has a role to play – from individuals, to industry, to government. There's not a silver bullet to solve our planet's problems, but there are hundreds of solutions – including limiting greenhouse gas emissions, protecting and restoring marine and coastal ecosystems, and strengthening scientific research. The awareness is there, now it's time to act. Our future depends on it!

# DR MICHAEL MANN

Climate change is expected to have large impacts on the weather systems around the world, changing where rainfalls occur and making storms more dangerous.

This is something that Dr Michael Mann has been researching. Dr Mann is a Distinguished Professor of Atmospheric Science at Penn State University, USA, and one of the world's most influential climate scientists. He's one of the scientists behind the Hockey Stick climate graph (page 66) and has played a large role in adding to the scientific understanding of climate change. Dr Mann is the author of more than 200 scientific publications, and has written 5 books (and counting!) about the climate crisis.

## Why does a changing climate impact the weather conditions we experience?

It is sometimes said that "climate is what you expect, while weather is what you get". Climate is just the longer-term statistics of the weather. That includes averages like the average temperature of the planet, which is increasing because of carbon pollution. But it also includes increases in "extreme events", like heat waves, droughts, floods, and superstorms, which are becoming more frequent and intense due to human-caused planetary warming.

## The term "global warming" makes it sound like the planet is just heating up, so why does climate change cause extreme cold weather as well?

Overall, globally, we are seeing a marked increase in hot extremes, as we would expect on a warming planet. However, there are some regions, for example in the USA and Europe, where we are seeing a tendency for larger swings in weather, including both heat and cold extremes, that might be tied to changes in the pattern of the jet stream as the planet warms up. There is still much scientific debate about the details of this effect.

## What impact will climate change have on weather systems around the world?

We expect more persistent weather extremes, in particular, heat waves, droughts, wildfires, floods, and superstorms as the planet warms up. We are actually now seeing these impacts play out in real time in the UK, the USA, and elsewhere. Examples from the summer of 2021 include the "heat dome" over the Pacific Northwest of the USA and the flooding along the US East Coast (including the state Pennsylvania where I live), which was linked with Hurricane Ida. In recent years we have seen the hottest temperature ever reliably recorded on the planet (54.4°C in Death Valley, California) as well as the hottest temperature ever recorded in Europe (48.8°C in Sicily).

# Interview with...

# DOMINIQUE PALMER

While climate change may impact the whole planet, some people will be more affected than others. This might be because of where they live or how much money they have. This concern with the unequal ways in which people are impacted by climate change is called climate justice.

Dominique Palmer is a climate justice activist from the UK, and is passionate about tackling the inequalities of climate change. She's part of the Fridays for Future organization, which works with young people around the UK that are pushing for climate action. Dominique has spoken at the UN's COP25 and COP26 conferences, and has been featured in news publications around the world.

# What is climate justice?

Climate justice is taking climate action for our planet while making sure that social issues are addressed, so that we can have a green and fairer future for everyone. It is about making sure that everyone has access to a healthy environment, with clean air to breathe, clean water, and green spaces. It concerns promoting good education for everyone, so that children have a better quality of life. Climate justice also looks at issues such as gender and race equality in order to promote a fairer world for everyone.

# Why are some people more affected by climate change than others?

Some people are more affected by climate change because of where they live and because they have less money. Some low-income countries in the southern hemisphere are already losing their homes and livelihoods because of climate change. For example, some communities who live by the sea have had to move from their homes because of rising sea levels.

# How can we make sure that everyone is included in global climate solutions?

Global leaders need to take action to ensure people's voices are heard and acted upon. Indigenous communities in particular are often not listened to, even though they offer solutions. Globally, Indigenous communities have protected 80% of our world's biodiversity. For example, our rainforests have been looked after by Indigenous communities for hundreds of years. Yet decisions made by global leaders have been destroying their environment and taking away their rights to protect biodiversity. It is essential that we include Indigenous peoples in the conversation about climate change.

# A question of
# DEGREES

The impacts of climate change are scary — and the more the planet warms up, the more extreme they will get.

Current global targets aim to keep global temperature rises under 2°C, with a focus on trying to keep them under 1.5°C. Otherwise, scientists warn that the climate crisis will cause catastrophic change to our planet. But while this 0.5 degree difference in temperature rise might not sound like a lot, the difference in the effect it has on the planet is massive.

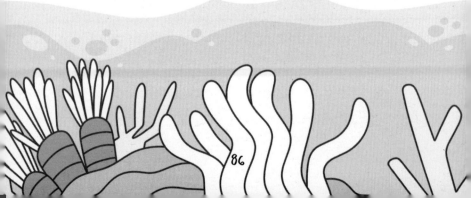

## Sea level rise

At 1.5°C, there would be 48 cm of sea level rise by 2100 (roughly double the sea level rise that's already happened), meaning 46 million people would be affected.

At 2°C, this would increase to 56 cm, impacting 49 million people.

## Sea ice

At 1.5°C, scientists predict the Arctic Ocean will have ice-free summers once every 100 years.

At 2°C, the Arctic Ocean could be ice-free in the summer once every 10 years.

## Extreme heat

At 1.5°C, 14% of the world's population could be at risk of experiencing at least one extreme heatwave every 5 years.

At 2°C, this increases to 37%.

## Coral reefs

At 1.5°C, 70–90% of the world's coral reefs could be lost by 2100.

At 2°C, almost all of the world's coral reefs would be lost by 2100.

## Water availability

At 1.5°C, 350 million people living in cities could experience extreme drought by 2100.

At 2°C, this rises to **410 million people**.

## Species loss

A temperature increase from 1.5 to 2°C could double or triple habitat loss for species.

## Extreme storms

More extreme storms will cause more flooding and rainfall in certain parts of the world. Changes in the ocean temperature and sea level will cause tropical storms to become more common each year.

## Food shortages

By 2100, a rise in temperature by 1.5 or 2°C would mean that there would be a shortage of crops like maize and wheat – particularly in places like Southeast Asia and South America.

## Health and poverty risks

Some people around the world are more at risk from the impacts of climate change than others – particularly Indigenous peoples, people who live by the sea, and communities that rely on farming. Climate change could mean these people lose their homes and livelihoods, causing an increase in poverty.

# 3

## Hope and
# SOLUTIONS

There's no denying that climate change is a big problem, and one that needs to be tackled now. When you read all the information about the impact climate change is having on the planet, it can be scary and you might be thinking that there's nothing we can do to make a difference.

But, that's where you're wrong! There is hope — it's not too late to slow down climate change (yet) and there are already some amazing solutions being developed to help humans reduce the amount of greenhouse gases we're releasing into the atmosphere.

From governments to scientists to individuals, we all have a role to play in tackling the climate crisis, and making sure that we protect the planet for our future and for other generations to come — whether that's through policy, technology, or small changes to our lives.

But where on Earth do we start? Many parts of our lives are having an impact on the climate, and it can be a bit overwhelming. That's why we're here to help! Chapter 3 is focused on looking at some of the solutions to the climate crisis – the international agreements in place, the eco-friendly alternatives to generating electricity, and the individual steps that we can take to combat climate change.

# The Paris
# CLIMATE AGREEMENT

Back in 2015, more than 190 countries came together at the Paris Climate Conference to sign an important document – the Paris Climate Agreement. It was the first binding agreement that brought countries around the world together to take action to combat climate change.

The Paris Climate Agreement is a part of the United Nations Framework Convention on Climate Change (UNFCCC). The UNFCCC was put together by world leaders in the 1980s, and it aims to reduce greenhouse gas emissions. Leaders meet every year to discuss the climate crisis. These meetings are called COPs – and it was at the 21st COP that the Paris Climate Agreement was created.

The Paris Climate Agreement aims to take more action on top of that which was already pledged by world leaders as part of previous climate agreements. Countries approved a range of targets for climate action, and have their plans, and any actions taken, reviewed.

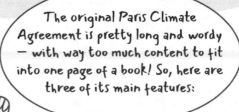

The original Paris Climate Agreement is pretty long and wordy – with way too much content to fit into one page of a book! So, here are three of its main features:

• Firstly, the Paris Climate Agreement aims to limit global temperature rise to a maximum of 2°C — although with the goal of keeping the rise below 1.5°C . To see why this is important, head back to page 88.

• Secondly, countries have their progress checked every five years, to make sure that they're still on track!

• Finally, the agreement calls for countries to support each other by sharing money and technology.

However, while the Paris Climate Agreement is a big step forwards in tackling climate change, some experts say that it's still not enough. They say that many countries aren't doing enough to meet the goals, that it doesn't do enough to punish countries that don't meet the goals, and that it fails to encourage countries to stop using fossil fuels.

And they might be right — since the Paris Agreement was signed, global $CO_2$ emissions have risen by 4% and continue to increase. That's why it's so important that we continue to put pressure on decision makers and politicians to stick to climate targets, and meet the aims of the Paris Climate Agreement.

# Carbon BUDGETS

You might have heard of a budget before — a budget with money is where you balance the amount you spend with the amount that you make, so that you don't get into debt by spending more than you earn.

A carbon budget is a similar concept, except the "money being spent" is $CO_2$ and the "money being made" is the global temperature rise.

A carbon budget is the amount of $CO_2$ we're able to release before we go over the maximum global temperature rise that we're aiming for. The Paris Climate Agreement states that this should be no more than a **2°C** rise. Scientists don't know the exact amount of $CO_2$ that can be released before we go over this target, but they can make a good estimate.

The UN's International Panel on Climate Change (IPCC) estimates that for at least a 66% chance of staying below the temperature rise target, humans must release no more than **1,000 gigatons of $CO_2$**. That number includes all the emissions that humans have emitted so far, as well as those to come in the future.

**1,000 gigatons of CO2 — what does that even mean?**
Instead of getting into the details of what this amount is physically, let's try to visualize it.

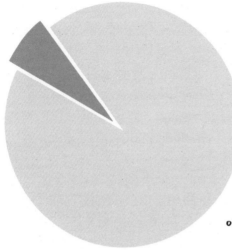

Imagine that this pie represents 1,000 gigatons of CO2. Once we've eaten the pie, we can't eat anymore. The problem is, this pie needs to last for all of humanity forever. And we've only got 130 gigatons of it left — just over one-tenth of the total!

## SO, US HUMANS NEED TO REDUCE OUR CO2 EMISSIONS TO PREVENT GOING OVER OUR BUDGET - AND FAST!

It's worth noting that carbon budgets are not perfect. It's hard to predict specifics like how long we have until we use them up, and they often ignore the fact that this CO2 is not released equally by all countries. China is currently the world's largest emitter of CO2, but the USA and Europe have been releasing CO2 for hundreds of years! But, what carbon budgets do do is give us something to aim for — which is useful for creating climate targets and policies.

# Clean ENERGY

As we found out in Chapter 1, the use of fossil fuels for
purposes like producing energy and fuelling our cars is
having a bad effect on the Earth's climate. If we can do
these important things without using fossil fuels, then we
can reduce the amount of emissions we release by a large
amount. Luckily, some very clever people around the world
have come up with ways that we can do just that – cut
down on our use of fossil fuels, and look to other, greener
methods of going about our lives.

Let's start with energy – after all, the use of fossil fuels for
the production of electricity around the world is the
leading cause of the climate crisis.

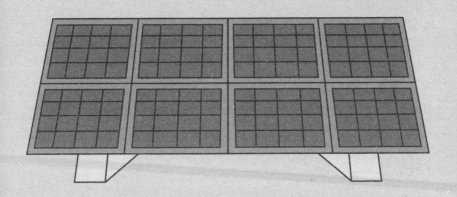

The good news is, there are methods of producing electricity that use other natural resources, such as the Sun and wind. This form of energy is called **renewable energy**, because it is produced from sources that nature can replace. Unlike fossil fuels – which take millions of years to form – the Sun, wind, and water are all replaceable in a short period of time. Renewable energy is also sometimes called **clean energy** because it doesn't pollute the air or water.

Sounds great to me!

# WIND power

One method of generating renewable electricity is by using the wind. Ever been out in the countryside or by the sea and seen rows of giant white towers with spinning blades? These are **wind turbines**, and, as you might guess from the name, they are turned by the wind to produce electricity.

## WIND ENERGY IS BECOMING MORE AND MORE POPULAR AROUND THE WORLD.

In 2021, wind energy helped to avoid over 1.1 billion tonnes of carbon emissions — equivalent to the total annual carbon emissions for the whole of South America!

## PROS:

- Wind turbines produce lots of electricity on a windy day

- They don't release greenhouse gases when generating electricity

- Turbines don't take up much space (for example, the land in between them can still be used for farming)

- They can be used to provide electricity in remote places

## CONS:

- Turbines don't produce much energy (or any!) on days when it's not windy

- Some people think that turbines are a bit ugly or noisy

- They can be harmful to wildlife, such as birds and bats

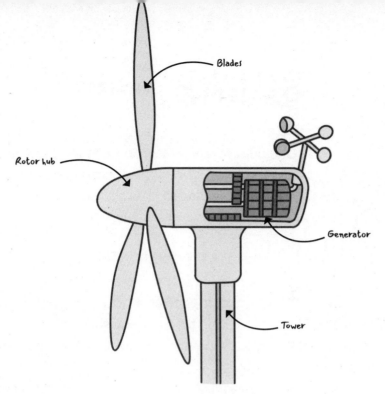

Blades

Rotor hub

Generator

Tower

## Blades

**Wind turbines** have huge blades on them. When the wind blows on the blades, it transfers energy to them, causing the blades to turn.

## Generator

The generator is a device that turns energy from the blades into **electricity**! When the blades of the turbine turn, they drive the generator, which produces electricity.

## Tower

Wind turbines are tall, allowing them to have very long blades. The larger the blades, the more wind they can capture. At the centre of the tower is a cable, which carries electricity down from the generator to the ground, where it can be connected to a power grid.

## Rotor hub

Also called a nacelle, this connects the blades to the generator, sometimes through a gearbox.

# SOLAR power

The Sun is a pretty incredible energy machine. Every hour, it beams more energy to Earth than is needed to meet the world's energy demand for a whole year!

This **SOLAR POWER** has huge potential to help humans generate energy renewably – that's where solar technologies come in! Through technology that can capture the Sun's energy, we can generate heat energy and produce electricity without having to use fossil fuels.

One of the main ways used to capture solar energy is through **SOLAR PANELS**. There are two types – ones that collect heat from the Sun (thermal panels), and ones that use the Sun's energy to produce electricity (photovoltaic panels).

Thermal panels work fairly simply – they use sunlight to heat up a fluid like water, which can then be supplied to a house or used to run a steam engine.

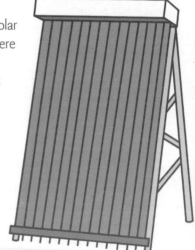

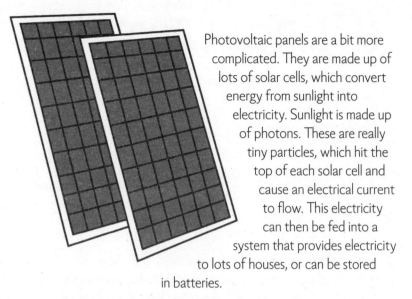

Photovoltaic panels are a bit more complicated. They are made up of lots of solar cells, which convert energy from sunlight into electricity. Sunlight is made up of photons. These are really tiny particles, which hit the top of each solar cell and cause an electrical current to flow. This electricity can then be fed into a system that provides electricity to lots of houses, or can be stored in batteries.

Solar panels can be set up in all sorts of places – on the ground, on the roofs of houses and buildings, and even on satellites!

## Pros:

- There is an endless supply of free sunlight to use!

- Solar panels don't release fossil fuels in the generation of heat or electricity

- They can be used to generate electricity in remote locations

## Cons:

- Solar panels can't be used at night, and are less effective on cloudy days

- They are relatively expensive

- Solar panels take up a lot of space

# HYDROELECTRIC
## power

It's fair to say water and electricity don't normally work well together – there's a reason we don't take our phones with us for a bath! But believe it or not, water can be an amazing way to generate energy renewably, and people have been doing this for thousands of years.

Hydroelectric power is power generated from the movement of water. Water is used to turn a set of rotating blades on a turbine (which is a bit like an underwater windmill). This is connected to a generator, which produces electricity when the turbine spins!

## BUT THE WATER NEEDS TO BE MOVING FOR HYDROELECTRIC POWER TO BE GENERATED...

... and there's a couple of different ways this can happen. The first is through reservoirs. Reservoirs are huge bodies of stored water, held in place by big walls called dams. When water is released from where it is stored in the reservoir, it is forced (by gravity and pressure) down pipes in the dam. The pipes lead the water to the turbine, causing it to spin. When less electricity needs to be produced (such as at night, when everyone is sleeping), the water can then be pumped back up into the dam so that it can be used again.

The other method is using the natural power that rivers can provide. Some of the river discharge can be diverted to turn a turbine.

## Pros:

- Hydroelectric power can be produced really quickly when needed
- It is reliable (if there is a good water supply)
- Reservoirs provide a water supply for local communities

## Cons:

- Building dams is expensive
- Building reservoirs involves flooding a valley, which can affect local wildlife
- If dams break, they can cause extreme flooding further downstream

# Other types of
# CLEAN ENERGY

Wind, solar, and hydropower are far from the only ways of generating electricity without using fossil fuels! Here are some other alternative forms of energy.

> Ever been to the beach, and had to move up as the seawater rises? These changes in water level are known as tides — they're controlled by the gravitational pull of the Moon!

## Tidal power

Tidal energy uses the tides to produce electricity. A **tidal barrage** (basically a massive dam) is built where the end of a river meets the ocean. When the tide comes in and out, changing the level of the water, it flows through a turbine, which causes electricity to be produced. It's pretty similar to hydroelectric power!

## Biomass

Biomass energy uses plants or animal material that
is burned to produce heat. Biomass energy
is renewable, as plants can be regrown in
a short period of time, but it does
still produce harmful gas
emissions (much less than
fossil fuels, however).

## Nuclear power

Nuclear energy is generated as heat in a process called fission, where
tiny particles called atoms are split. The energy is used to heat water
to create steam, which turns a turbine. A massive amount of nuclear
energy can be released from just a small amount of atoms.

But this energy comes at a cost. Nuclear power stations become less
efficient over time and eventually must be closed down. They also
produce very toxic, radioactive waste which remains dangerous for
thousands of years. So the use of nuclear power stations to generate
energy is hotly debated.

# ALTERNATIVE
## transport

Green transport doesn't just mean painting vehicles green – although, that would be pretty snazzy! What it actually means is developing methods of transport that use less (or no) fossil fuels to power them. Here are some examples:

## Electric vehicles

Electric vehicles are powered by (you guessed it) electricity! They store electricity in a battery, which is used instead of petrol or diesel in an engine.

There are a couple of types of electric vehicles – hybrids and pure-electric. Hybrid vehicles use petrol or diesel as fuel alongside electricity from a battery, which is recharged using energy from braking or by plugging them in (a bit like charging a phone!). Pure-electric cars run only on electric batteries.

## Trains

High-speed rail networks are popping up around the world – this is where trains travel on special tracks at a higher speed to normal trains. High-speed trains are great, as they use little or no fossil fuels to run (the trains are powered by electricity) and allow people to travel really quickly. This means they can encourage people to use cars less, or even avoid flying!

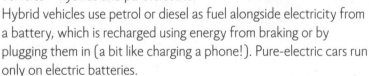

# Electric planes

In 2016, a plane made history
by being the first to fly around the
world using no fossil fuels – instead, it
was powered by solar energy! Electric planes
are a great way for the aviation sector to reduce
greenhouse gas emissions in the future, but they've still got
a long way to go in making sure that the planes can travel long
distances and carry lots of passengers.

There are drawbacks, however. Electric vehicles are expensive and can only travel a short way before they have to be recharged. And they're only fossil-fuel free if the electricity that powers them is renewable!

Also, the batteries used in electric vehicles are made from precious metals often sourced from South America and Central Africa, where the mining of the material consumes local water supplies and forces people to work in dangerous conditions. So we need to improve how batteries are made and used.

# Sustainable transport
# WORLDWIDE

## Helsinki, Finland

77% of journeys made in the city of Helsinki are done via walking, cycling, or public transport, and using these methods of transport means you can also get free access to one of Helsinki's saunas! The city also has 1,200 km of cycle routes.

## Curitiba, Brazil

Curitiba, a city in Brazil, led the way in developing its Bus Rapid Transit (BRT) system, an advanced bus network that is more reliable and holds more people than conventional bus systems. Brazil still has the highest use of BRT systems in South America, an estimated 10.7 million daily riders.

# Seoul, South Korea

Seoul's Cheonggyecheon river was concreted over in the 1970s to build a highway. In the early 2000s, work began to restore the river – the road was demolished, and to make up for the loss of transport, Seoul expanded its BRT system and improved its systems for non-motorized transport. Since then, there has been a rise in bus and subway use in Seoul, an increase in biodiversity by 639%, and the river even provides flood protection for the city.

# Zurich, Switzerland

In 2017, Zurich was ranked as the second-best city for sustainable public transport! 57% of people who live in Zurich use public transport as their main way of getting around, and there are also 2,250 bikes available to be rented around the city. Plus, the largest train operator in Switzerland is powered by 90% hydropower!

# Tokyo, Japan

Tokyo is the most populated city in the world, and it has some great transport networks to help its inhabitants get around. Its subway system is used by an average of over 7 million people every day! Tokyo is also connected to other cities in Japan by Shinkansen, the world famous bullet trains powered by electricity that can reach 320 km per hour!

# Carbon
# OFFSETTING

Our carbon footprint is the amount of carbon dioxide we're responsible for releasing. To combat climate change, many of us — from individuals to big businesses — are trying to reduce our carbon footprint, or at least make up for it.

One way of doing that is through carbon offsetting. Carbon offsetting is where you make up for the carbon dioxide you release by **PREVENTING** carbon from being released somewhere else, or by taking the same amount of carbon dioxide back out of the atmosphere.

The aim is that the amount of carbon dioxide in the atmosphere remains at the same level. So, say you release one tonne of $CO_2$ into the atmosphere in a year — to offset that amount, you need to ensure that one tonne of $CO_2$ is made up for by:

- **PREVENTING** this amount of $CO_2$ from being released somewhere else

- **REMOVING** this amount of $CO_2$ from the atmosphere and storing it

Preventing $CO_2$ from being released is called "carbon avoidance". Carbon-avoidance projects are things like building lots of wind turbines to reduce emissions from fossil fuels — they don't affect the CO2 that's already in the atmosphere, but they do prevent an equal amount from getting up there. It's like swapping the emissions from one place to another.

Unlike carbon avoidance, **CARBON REMOVAL** has an impact on the CO2 already in the atmosphere. If one tonne of CO2 is released, it is offset by taking one tonne out of the atmosphere through natural methods like growing more trees, or through technological methods like removing carbon dioxide from the air and storing it deep underground. Sometimes this carbon dioxide can be captured as soon as it is produced, such as at fossil fuel power plants.

Carbon offsetting is often used by businesses as an important part of reaching carbon neutrality, where businesses don't add any extra carbon dioxide into the atmosphere. But carbon offsetting alone isn't the solution to being carbon neutral — the first step is to reduce carbon emissions as much as possible, and then offset the little that remains.

Imagine carbon removal is like a leaky bucket, where the bucket is the atmosphere, water is CO2, and carbon removal is the holes in the bucket. If you poured water into the bucket at the same rate that it was flowing out, the water level would remain the same! That's how carbon removal works.

Some people worry that too much emphasis is put on carbon offsetting and that it takes away from the real solution to the climate crisis — cutting carbon emissions. Plus, carbon offsetting projects can have a negative impact on local communities, or even damage the local environment.

# What are
## countries DOING?

We've already seen that some countries are having a pretty negative impact on the climate. But the good news is, many countries are already starting to **TAKE ACTION** to protect our planet. Here are a few countries that are leading the way...

## Costa Rica

Since 2019, **99%** of the energy generated in Costa Rica has been renewable, and the energy is accessible to 100% of the households in the country. Because of this, in 2017 Costa Rica set the record for the most days running only on renewable electricity – an **AMAZING 300 DAYS!** Costa Rica is also a popular tourism destination, yet has dedicated **26%** of its land to wildlife reserves and national parks, meaning it has been able to balance the economic interests of tourism with the protection of Costa Rica's incredible biodiversity.

## Denmark

In only the last few decades, Denmark has halved its emissions while also doubling its economy! And it's not stopping there – Denmark has also pledged to cut its emissions by another **70%** by 2030, with many areas of the country already setting out plans of how to reach this target (such as through more public transport, electric buses, and renewable energy).

## Morocco

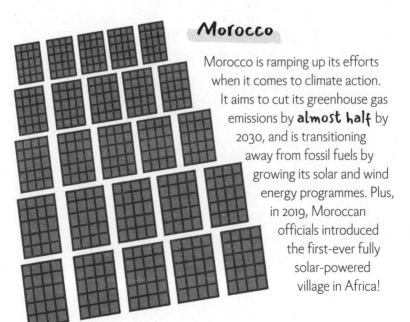

Morocco is ramping up its efforts when it comes to climate action. It aims to cut its greenhouse gas emissions by **almost half** by 2030, and is transitioning away from fossil fuels by growing its solar and wind energy programmes. Plus, in 2019, Moroccan officials introduced the first-ever fully solar-powered village in Africa!

# What are other countries around the world up to when it comes to climate action?

## CHILE

Chile is one of the leading countries promoting climate policies in South America. It is meeting its targets for increasing renewable energy production, and was one of the leading nations in launching the Climate Ambition Alliance (which commits countries to becoming carbon neutral by 2050).

# UK

The UK is ranked as one of the best countries in the world when it comes to climate action. It was the first nation in the world to legally agree to reaching net zero carbon emissions by 2050 and has pledged to cut carbon emissions by **68%** by **2030**. It is also the largest producer of offshore wind power (energy produced from wind turbines in the ocean) in the world!

# SWEDEN

Sweden has some ambitious targets! It aims to reach **NET ZERO** by 2045 (which it could even achieve by 2030). More than half of Sweden's energy is renewable, and it aims for its transport sector to be fossil-fuel free by 2030.

# INDIA

India is still responsible for a large portion of global emissions (mainly due to its reliance on coal), but it is taking steps to address this. It is currently on track to meet its 2030 emissions goal, and has targets to increase renewable energy generation and electric vehicle use.

# NEPAL

By 2030, Nepal aims to increase the sales of electric vehicles and have one-quarter of households using electric stoves. Through this, Nepal hopes to reduce almost a third of transport emissions and more than a fifth of cooking emissions. Nepal has also been praised for its inclusion of women, Indigenous peoples, and youth in its climate action plans.

Even though lots of countries are recognizing that climate change is a problem, many of them are still not taking action that is radical or urgent enough to keep global warming to below 1.5°C. The nations taking the most extreme action to combat the climate crisis are often those at the most risk from its consequences, while also being the least responsible for the problem in the first place.

These countries also need funding to help with their climate actions. So it's important that wealthy countries include "climate finance" as part of their policy — money loaned to less wealthy countries to help them adapt to the consequences of climate change.

It's also key that these targets aren't just targets — it's super important that governments actually take the action required to meet them.

# FAKE NEWS

## CHINA'S THE ONE TO BLAME!

Ah, the classic tactic of shifting the blame completely onto someone else – in this case, a different country. Yes, some countries are responsible for more carbon emissions than others, and China does release a lot of those emissions. But so do other countries, such as the USA, Russia, and India. Plus, to point the finger as an excuse for not doing something yourself is unhelpful at this point. To tackle climate change with the urgency it needs, we all have to work together globally, share resources and information, and take action while we still can.

# Eco-ANXIETY

The effects of climate change can be worrying. We see the scary facts about the rate at which the climate is changing, and the effects that it will have on our lives and for all future generations to come. Sometimes, this worry can make us feel anxious or upset. This is called **"ECO-ANXIETY"** and it is a fear of the future of our planet, often relating to the impacts of climate change.

Eco-anxiety is experienced by people young and old, but it's particularly felt by people who will be impacted the most by climate change – such as Indigenous communities, and people who rely on the environment for food and shelter.

That's why eco-anxiety is really common among us kids. One of the reasons eco-anxiety is felt by younger generations is because climate change is something that will affect our lives, and it can feel like we can do nothing to stop it – like it's out of our control. Surveys have found that over two-thirds of young people experience eco-anxiety, so if it's something that you've felt, you're definitely not alone.

So how can we combat it? Often, the best way to feel less anxious about the environment is to **TAKE ACTION YOURSELF** to take back some control. This could be through:

- Talking to someone about your anxiety, like a parent or trusted adult

- Learning more about climate change and spreading the word

- Making small changes to your life to be more eco-friendly

- Spending time outdoors and enjoying nature

- Joining an organization that works to address climate change

Really, there's no right or wrong way to combat eco-anxiety — the best thing to do is find a method that works for you.

While tackling climate change means that governments and companies need to act, making changes yourself can help you to feel less powerless.

Turn over to the next page for some top tips on tackling eco-anxiety!

## Interview with... KATIE THISTLETON

Katie Thistleton is a presenter, journalist, and author – we grew up watching her on TV, when she presented on the UK kids' channel CBBC. Since then, she's gone on to host podcasts, radio programmes, and documentaries. Katie is also passionate about mental health – she's training to be a counsellor, and is an ambassador for numerous mental health charities.

### Why does climate change make us feel anxious?

Climate change can make us feel anxious because we hear lots of stories about what may be in store for our planet, and it's natural for us to be concerned about our futures and our world. A certain amount of this concern is helpful, but sometimes anxiety can start to happen more frequently than we would like.

## Will eco-anxiety go away?

It is likely there will always be something which causes concern when it comes to the world around us. Anxiety comes in many forms and it's likely you will feel anxious about lots of different things throughout your life. The best thing to do is learn how to deal with your anxiety, as you can't always be in control of the things you worry about.

## What are your top tips for overcoming eco-anxiety?

Take action. Eco-anxiety is particularly difficult because it can feel as though we are helpless, and that we won't be able to change the world on our own. However, the actions of each human being are very important, and just spreading the right messages to the people you know can make a real difference. Doing what you can, even if it feels small, can make you feel as though you are doing your bit, which may ease that anxiety.

Accept that you have eco-anxiety. It's perfectly natural to feel anxious, and it is something we all experience now and again. Talk about your worries with friends or family members, many of them may feel the same way, too!

Protect yourself from catastrophizing. We are not our thoughts, and sometimes our thoughts can get out of control and make us worry about things we don't need to worry about. If you find your eco-anxiety is really affecting your day-to-day life, it might also be worth trying not to read or watch too much news, as sometimes that can also be over-dramatized.

## Who should we talk to if we're feeling anxious?

A trusted adult such as a parent, carer, or teacher is always a great person to talk to when you're feeling anxious. Some of your friends may also feel the same as you and it might make you feel better to chat to them and know you're not the only one – you're definitely not! Your GP can also help if you are struggling, as can helplines such as Childline – you can call them on 0800 11 11.

# How we can make
# CHANGES

While it's super important for governments and businesses to take action to combat the climate crisis, there are things that we can all do to help. After all, none of us like the idea of sitting around and waiting for those in power to do the work – and we might be waiting a long time!

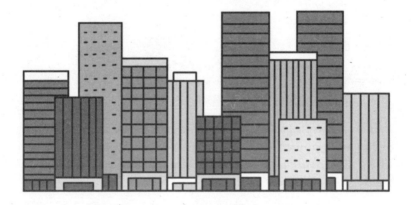

We can all play a role in reducing the impact our daily lives have on the planet. The way we live in the **GLOBAL NORTH** (the richest and most industrialized countries on Earth, mainly those in Europe, North America, and some parts of Asia) is causing the emission of lots of carbon dioxide. By making changes to our own lives, we can reduce the amount of $CO_2$ we're responsible for.

123

# How to use less
# ENERGY

Cutting back on the amount of energy we use might sound complicated, but even little changes can help us to reduce our **CARBON EMISSIONS** and save money on fuel bills!

Here are some simple ways you can start reducing the energy used in your home:

### Try the five-minute shower challenge

Love your shower karaoke sessions? So do we. And whilst we don't want to put a dampener on the performance, it's worth choosing a shorter song – ideally, one less than five minutes long! Hot water used in showers requires energy to heat and pump it, so by shortening your time washing, you can reduce the energy used and save money on your water bill.

Sounds like a win–win to me! Showers are still better for the environment than bathing – the Environment Agency estimates that a five-minute shower uses one-third of the water used for a bath!

## Switch off

Make sure you turn off all your lights when leaving a room. Will this really have any impact, you might ask? It might appear a very small change, but a 2018 study found that around 10% of the UK population regularly leave their lights on in rooms that they're not in — the carbon emissions from this is equal to 62 flights every day! Not such a bright idea.

## Light-bulb moment

Swap your light bulbs for LEDs or other low-energy lights. LEDs convert **95%** of the electricity that they use into light, with only 5% being wasted as heat. While they may be a little more expensive to buy, they can also last up to 25 times longer than traditional light bulbs — so, they're cheaper in the long run!

## Layer up

Make sure to stop the heat from escaping from your house in the first place! Put aluminium foil behind your radiators to reflect heat back into the house and try putting draught stoppers underneath your doors.

My Dad always tells me to "put on a jumper" if I'm cold. As annoying as that might be, it helps the planet — putting on layers is an easy way to reduce your carbon footprint!

# Switching to
# RENEWABLES

If you really want to get serious about the $CO_2$ that comes from your home energy use, then it's time to look at **INSULATION**. This is the material that is used to slow the rate at which heat is lost. It's like putting a tea cosy on a

teapot – the tea cosy insulates the teapot, keeping the tea inside it warmer for longer! In houses and flats, insulation serves a similar purpose – it slows down the escape of heat from your home, by filling in spaces where heat could be lost, such as in the space between walls. This is crucial for reducing energy usage. However, even with insulation, homes still need to be heated somehow! Switching to a renewable energy provider so the energy supplied comes from sources like wind (page 98) or solar (page 100) could be the answer.

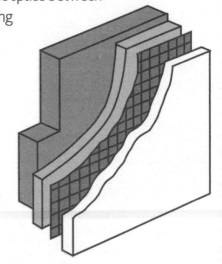

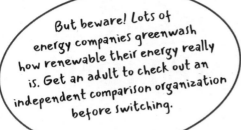

But beware! Lots of energy companies greenwash how renewable their energy really is. Get an adult to check out an independent comparison organization before switching.

Don't worry if you can't switch to renewable energy straight away — it can be more expensive than fossil fuels. Small changes around your home are a great way to start.

# Reducing MEAT CONSUMPTION

Growing and raising farm animals has a really big impact on the planet. So, by cutting down on the amount of animal products we eat and drink, we can also reduce the amount of greenhouse gases we're responsible for releasing!

Ever heard of **VEGANISM**? It means not using anything that comes from animals. Vegans don't eat meat or dairy (products such as milk and cheese), or wear clothes made from leather or wool.

Veganism is not new – it can be traced back to India in the 5th century BCE!

Being vegan can cut down an individual's carbon footprint by up to **73%** – so it's sometimes described as the **single biggest way** to reduce your impact on the planet. This is because foods needed for a vegan diet tend to need much less water, land, and energy than animal products.

## BUT, THAT'S A BIG STEP TO TAKE!

Becoming vegan can be difficult, if not impossible. Vegan alternatives can be more expensive, and some medical conditions make being vegan impractical. If you do decide to become vegan, it's vital you get the right amount of vitamins and minerals to stay healthy. The good news is that there are lots of tasty foods that are great alternatives to meat, such as tofu, beans, and vegetables. Alternatively, just cutting down on meat is something most of us can do. Avoiding beef one day per week, for example, can save the same amount of emissions as driving **560 KM** in a car! And, one in five people who do this end up becoming vegetarian after five years anyway!

Veganism isn't a one-size-fits-all diet, however. In some Indigenous communities, hunting and eating animals is a necessity, and is sustainable.

129

# SHREYAA & ESHA VENKAT

If food waste were a country, it would be the third-largest emitter of greenhouse gases in the world (behind China and the USA). Yet, reversing current food waste trends could save enough food to feed **2 billion** people! Shreyaa and Esha Venkat are on a mission to combat food waste and hunger. They're sisters from the USA, and founders of a non-profit organization called NEST4US.

## 1. How is climate change linked to food waste?

When you waste food, you're not only losing the actual food, but also the energy consumed from labour and machinery, the biodiversity lost from deforestation, and the time spent to grow, transport, and sell the food. Also, food waste gets taken to landfills, where it causes a **LOT** of environmental damage. Food production accounts for around 25% of global greenhouse gas emissions, and landfills account for over half of that percentage – around 15% or 136 million tonnes per year.

## 2. Why is so much food wasted?

Even before food reaches the shops, wastage happens. This is mainly because of bad weather, processing issues, unstable markets, and overproduction. Once food is in the shops, food might be wasted because of overbuying, poor planning, and issues around food safety.

## 3. What are your top tips for reducing food waste?

1. Educate yourself on these issues and start taking measures to stop food waste. A good place to start is by taking the **Food Waste Pledge** on the NEST4US website (www.nest4us.org), which lists simple ways to start reducing the amount of food you waste every day!

2. When shopping, make a buying plan and only get what you need.

3. Donate your excess food to a local homeless shelter, food bank, or nonprofit organization near you.

4. See if you can organize a Share Table at your local school. This involves a table where students can drop off food that they may not have had time to eat, such as an unopened milk carton or granola bar.

5. Simply because of their appearance, businesses and even farms are forced to throw lots of "bad-looking" vegetables away. So a great way to help reduce food waste is by purchasing these "imperfect" foods from those shops that do sell them!

6. Use online platforms to promote your views. Through our **#FightHungerFriday** social media campaign we show how people can help stop food waste and feed the hungry every Friday.

7. Join or host your own sustainability workshop! Through our **Innovation Across Nations** workshop series in collaboration with Kids Against Plastic, we promote climate action by connecting people worldwide.

8. Lastly, you can join our volunteer family, of course! Check out our website www.nest4us.org to fight food waste alongside us.

# Reducing
# FOOD MILES

> It would be so cool if we could grow fresh food in our back garden or windowsills! That would save me lugging home all those heavy shopping bags of fruit and vegetables (talk about a workout!).

> While that's a lovely idea, different weather conditions and eating habits mean that many of us rely on buying most of our food from shops.

The **food miles** of the things we eat (that means, the distance they've travelled from where they're grown to where they're eaten) can be pretty bad for the climate. This is especially true in Western societies, where our eating habits have changed over the years from seasonal food that's readily available, to "exotic" foods from all around the world. So, just getting dinner on our plate can result in a lot of carbon emissions.

Foods that rot or go out-of-date over a short period of time, such as bananas or avocados, have to be transported to places very quickly, often by aircraft. Flying in food can release 10 times more carbon emissions than food transported on roads, or about **50 TIMES** more emissions than shipping! And even when food has reached our own country, it may need to be transported hundreds of kilometres to reach our doorsteps or local shops.

It doesn't stop there. Foods grown in our own countries in artificially heated greenhouses can even produce more emissions than those flown in from abroad!

## SO, WHAT CAN WE DO TO REDUCE THE EMISSIONS FROM THE FOOD WE EAT?

**1.** Shop **locally**. Food grown near to where you live hasn't had to be transported huge distances to reach your plate, and it can give a boost to local, independent farmers in your community.

**2.** Buy **seasonally.** Fruits and vegetables can often only be grown at certain times of the year. By only buying this produce when it's in season, it means it won't have been flown across the world, or been grown in artificial environments that need lots of energy.

**3.** Plan ahead. If you live in a country like the UK, it's a struggle in the winter months to buy enough seasonal produce to be healthy. So, freezing fresh, summer fruits and vegetables can mean you have a nice stock to keep you going through the winter, in the form of soups or smoothies. Certain fruits and vegetables need to be frozen in different ways, so time to crack open a cookbook and get thinking!

# HARRY WATERS

Getting your family and friends involved with your action to be more **ECO-FRIENDLY** can be a great way to cut down your impact on the planet at home. But, it's not always easy – sometimes they might not be interested or supportive of the changes that you want to make, which can be really frustrating.

Luckily, real-life parent and eco-warrior Harry Waters is here to give us a hand. Harry is the founder of Renewable English, through which he runs free online classes, creates resources, and hosts podcasts all about climate change! He and his daughter Ali are also keen **LITTER-PICKERS**, and as a family they are working hard to be as eco-friendly as they can be.

## My parents don't think climate change is a big problem — how can I change their minds?

Changing an adult's mind is never an easy task. You have to be persistent. Your parents only want the best for you, so showing them that making a difference to the climate crisis is the best way for you to have a better future is a good way to convince them. Talk about what you've learnt at school and draw their attention to the growing attention in the media. It won't be long before they realize it is a big problem, and one that needs action immediately.

## What are simple ways to get my family involved in sustainability?

Suggest those easy changes like using your own bags at the shops, not taking the car for short journeys, and using a **REFILLABLE BOTTLE**. Start a litter-pick, and why not try some easy, plant-based recipes.

## Do you have to be a certain age to make a difference?

There is no age limit — you can neither be too young nor too old! My daughter has been passionate about the planet since she was 6 and my Nan is still passionate at 86. It's never too late and it's certainly never too early. How do you think tackling climate change can be made into something appealing? Find your **PASSION**. You can't take on everything at once, so find what inspires you most and build from there. When you're passionate about something it'll become far easier to motivate yourself to keep going. Your passion will also bring others with you on your journey. My passion is **EDUCATION**, so I try to involve a discussion of the planet in all my classes. I want to build a love for the Earth in my students. The more you learn, the scarier things may become. But be brave and be passionate. There are many negative outcomes to doing nothing, you can only have a positive effect by doing something.

# Clean
# TRANSPORT

Road passenger transport accounts for around **45%** of the total carbon emissions that come from transport. But luckily there are things we can do to reduce this, and some of the methods are not only good for the planet, but also for us!

Switching away from fossil-fuelled transport is a great way to reduce greenhouse gas emissions. Electric cars powered by renewable energy are responsible for very little emissions compared to cars that are fuelled by petrol or diesel. But for some people, this can be an expensive and impractical step to take – and as we saw earlier in the book, **electric vehicles aren't without their flaws!**

Changing to a cleaner car isn't all we should be considering... we can also look at alternatives to travelling by car, especially for shorter journeys.

In the UK, 60% of journeys made by car are between 1.6 km–3 km. The cleanest way to cover this distance is to walk or cycle, with zero carbon emissions! This not only has a positive impact on emissions and road traffic, but also helps us get our daily exercise.

When walking or cycling isn't possible, public transport is a great alternative. A large amount of trams, buses, and trains now run on renewable or electric sources – just the ticket *(ba-dum-tsh!)* to low-emission transport.

AND, KEEP YOUR EYE OUT FOR THE HIREABLE ELECTRIC SCOOTERS AND BIKES WHIZZING AROUND CITY CENTRES!

By creating new habits and making small positive changes to the way we travel, we can live more sustainably and maybe even get fitter in the process!

# Tackling
# OVERCONSUMPTION

Overconsumption is one of the key drivers of the climate crisis, but what does it mean?

When more resources are used than are needed, and at a faster rate than they are produced, this is called overconsumption. It's like buying 10 new T-shirts when you already have lots, or only need one new one. It doesn't make sense, and it's not good for the planet – after all, each of those T-shirts would need fabric to make them from, people or machines to put them together, a lorry or plane to transport them to different countries, and a shop to sell them in. **That's a lot of energy and emissions!**

Overconsumption is pretty easy to fix on an individual level — buy less!

OK...
But, it can be harder than it seems, especially as most of the shops we buy from and adverts on TV are pushing messages to buy more, more, more!

Here are some **tips** for reducing your overconsumption:

## Think before you buy

Do you really need that new thing? How long will you use it for? How long is it designed to last? These are all important questions to ask yourself before buying something new. By carefully thinking about the things we buy, we can cut down on the amount of unnecessary items produced and reduce waste. It also makes the new things that you do get even more special!

## Steer clear of single-use

Single-use items – ones that are used once before being chucked in the bin – are a perfect example of overconsumption. They mean that more and more  items need to be made just to replace ones that were used the day before. Disposable plastics are perfect examples of this – things like plastic bottles are only used once before we get rid of them. So, cutting down on the amount of single-use plastics you use is a great way to help the planet. Try refilling a **REUSABLE** alternative instead, which you can use again and again and again and again...

## Give the things you own a value

Imagine your favourite toy or teddy bear. You wouldn't want to throw it against a wall or drag it through mud. If we care about something, we look after it, which means it's more likely to last a long time. Try looking after the other things you have in the same way.

But overconsumption is a bigger problem than our individual choices. We need to make sure that businesses are designing products that last for a long time, and can be repaired instead of replaced. Check out Chapter 4 for tips on how to use your voice to get corporations to stop driving overconsumption, too!

# Becoming
# NET ZERO

We've already talked a bit about net zero earlier.

With net zero, you don't contribute to global warming — the balance of greenhouse gases in the atmosphere is kept the same through reducing your emissions.

Being net zero can be applied to businesses and governments, and also us as individuals.

It can be a huge thing to achieve! Here are some tips to taking your first steps towards a net-zero life...

# Calculate your carbon footprint

The best place to start is by working out the amount of carbon emissions you or your family are currently releasing. There are lots of carbon footprint calculators online that can help you to do this for free — you just need to put in some of your around-the-house habits, details like the amount you travel, and the calculators will give you an estimate of your yearly emissions broken down by section! Pretty handy.

## Take it section by section

Once you've calculated your emissions for different parts of your life, you can work on addressing them. Don't think you have to do it all at once — for example, if most of your emissions come from your energy usage, start by trying to cut that down.

## Carbon offset

The best thing to do is to reduce your carbon emissions as much as possible — but if there are some unavoidable ones that have slipped through the net, that's where carbon offsetters come in. Carbon offsetting alone isn't the way to reach net zero, but reliable carbon offsetting schemes can help you to limit your impact on the planet by tackling those emissions that you didn't manage to avoid.

## Don't worry if you can't reach it

But going completely net zero in your own life is not easy **(OR OFTEN, CHEAP)**, so don't beat yourself up if you can't take all the steps to get there. After all, if it was super easy to balance greenhouse gas emissions, surely we'd all be living net zero lives already! The important thing is to start doing something to reduce your emissions, even if it is just something small.

# 4

# We can make a
# DIFFERENCE!

We might not be old enough to vote, to run a business, or even drive a sustainable car – but that doesn't mean us kids don't have an important voice and role to play.

In this chapter, let's look at how young people can use our voices to push for climate action!

143

# On this DAY

Action against climate change has been happening for decades! Let's look at some of the most memorable moments of climate activism through the years.

## 22ND APRIL, 1970

The world's first-ever Earth Day took place, to raise awareness of pollution. Earth Days still take place every year!

## 11TH JUNE, 1992

Canadian environmentalist Severn Cullis-Suzuki gave a speech at the UN's Rio Summit, at the age of **12!** It was described by the former US Vice President Al Gore as **"the best speech [he] heard all week!"**

## 12TH DECEMBER, 2009

Alongside the UN's COP15 Summit in Copenhagen, 40,000–100,000 people attended a march to call for global agreements to tackle climate change.

## 21ST SEPTEMBER, 2014

The People's Climate March took place in New York, attracting 400,000 activists to take part.

## 20TH AUGUST, 2018

Swedish activist Greta Thunberg held her first strike outside the Swedish parliament, inviting young people around the world to join her. She held a sign saying "Skolstrejk för Klimatet" – Swedish for "School Strike for Climate".

## 20TH SEPTEMBER, 2019

The Earth Strike – the biggest climate strike ever – took place in over 150 countries around the world, with an estimated **4 MILLION** people taking part. The strike, led by Greta Thunberg, was part of the Fridays for Future movement.

## 11TH OCTOBER, 2021

Protesters, led by Indigenous activists, gathered outside the White House to demand the US President Joe Biden stop allowing fossil-fuel projects to go ahead, and to declare a state of climate emergency.

# Kids making
# A DIFFERENCE

You're never too young to make a change. Activism can be whatever you make it – it doesn't have to be big, bold actions.

# HERE'S WHAT SOME OF OUR KAP CLUB MEMBERS HAVE BEEN UP TO!

## INAAYA

Inaaya is a keen environmentalist and has presented to her local politician ideas about how her neighbourhood area can be greener. Inaaya has appeared on TV to speak about her action, and she hosts her own radio program every two weeks. Inaaya is on a mission to make her school plastic-free and has already managed to get her school to get rid of plastic water bottles.

## SKYE

Skye started her own campaign at the age of 10 years old. She's taking action against plastic toys on magazines, and her petition has had over 60,000 signatures! Skye's already had success with her campaign – a UK supermarket and a magazine subscription have already agreed to ditch plastic toys.

## ALFIE

Alfie's 2021 New Year's resolution was an exciting one – he and his family set up Be A Planet Superstar, a "call to action" campaign that celebrates those who are taking action to help the planet. Alfie has spoken on the radio, launched his own school challenge, and helped to plant thousands of trees.

## THOMAS

Thomas has been working against plastic pollution since the age of 6! Aside from his own clean-ups (including his local river!), Thomas shares his environmental message in creative ways. He entered his local fancy dress competition in a costume he made from plastic rubbish, with the slogan "the sea is not a dustbin", and went on to win it! He's also a keen member of the Scouts organization and is working hard to make his troop more "plastic clever".

# Interview with...

# HAMANGAI PATAXÓ

Hamangai is from Brazil, and belongs to the Pataxó Hã-Hã-Hãe Indigenous peoples. She is a passionate activist for the protection of nature and the rights of Indigenous peoples. Hamangai attended COP24 in Poland in 2018, spoke at the Villagio per la Terra (Village for the Earth) in Rome about the Amazon and the importance of protecting the people who live there, and was part of the 2019 Young Activists Summit at the UN Headquarters in Geneva. She is now studying veterinary medicine at university.

Hamangai grew up in a part of her village where there was no drinking water. When she was older, she and her parents moved to another area of the village where there was a beautiful river full of wildlife. But since then, the climate has changed, meaning the river she grew up next to is no longer the same. Her dream is to one day see her village with its native forests grown back and its rivers flowing again, but she says that there's a long road ahead before this happens.

### How have you seen climate change affecting your community?

The rivers and many animals are disappearing. This is really sad, because we need water to survive.

### What action do you want to see from governments to better protect the natural world?

I would like the government to listen to us Indigenous peoples, and to truly commit to taking care of nature, as we all should be doing. I don't want to keep hearing lies any more – I want to see actions.

### What more needs to be done to support the work of Indigenous activists like yourselves and your communities?

We need to take care of those who look after forests and animals. We need to fight so that everyone is heard. We need our lands to be in our own hands, and not in the hands of those who will destroy them.

### What message would you give to any young person reading this book?

You are seeds; we need to spread ideas and dreams with courage and determination for a better world for everyone. Feel the nature, step with your feet on the soil, feel the rain, and embrace the Indigenous peoples' fight as your own.

**"I DREAM OF SEEING MY VILLAGE WITH ITS NATIVE FORESTS GROWN BACK AND ITS RIVERS FLOWING AGAIN, BUT THERE'S A LONG ROAD AHEAD BEFORE THIS HAPPENS."**

# Interview with... DEV SHARMA

Dev is the Youth MP for Leicestershire in the UK, representing over 230,000 young people in the UK Youth Parliament. Alongside his political career, Dev is a passionate campaigner against food poverty, and an advocate for all young people to have equal access to a good diet. He's chairperson of the youth-led movement Bite Back 2030, an ambassador for the Food Foundation, and in 2021, he was successful in getting the UK government to ban junk food advertising.

### Why do you think it's important that young people are given the chance to lead decision-making processes?

Young people aren't just the leaders of tomorrow. Collectively, they have the energy, skills, and ideas to improve society and the environment today. Despite this, society often fails to listen to young people's views, or recognize their ability to make a positive difference. That's why movements like Bite Back 2030 exist together with young people to achieve their shared vision.

## How can young people's voices be better listened to?

It is important to equip young people with the tools and the platform to campaign for change – especially when talking about sensitive topics like food poverty or climate change, which can be easily misrepresented by politicians and in the press. Creating campaign groups to get our voice heard is really impactful. These help maintain a youth-presence at a policy level and help ensure representation isn't tokenistic.

## As a youth MP yourself, what role do you think politics should play in tackling the climate crisis?

Politics is of immense importance, especially now, because we know individuals cannot solve the climate crisis. Governments can, and they need to step up and step in. Climate change is the toughest political issue we've ever had to face as a society. However, the politics around it seems to lack the urgency felt by many of us. This is why we have a responsibility to make the conversation a lot more urgent. We need to demand that politicians make climate change a priority.

## What is your message to a young person wanting to become an activist?

No one is expecting you to be an expert straight away. You will learn and grow. Remember, activism isn't linear. It is a growth journey, and you will find new things that will test and challenge your thoughts along the way. Your passion and voice are enough. People want to hear why you are interested and passionate. And that is what matters.

When campaigning, it can often feel like you are speaking into an empty echo chamber. Remember though that others are listening, but they just might not be replying. If you stay true to your mission and persist with true passion and strength, there is no one stopping you. Remember that you are not alone in the fight. Power comes in numbers. So team up, mobilize, and speak up. Humans thrive when they work together within a community. Each talent, skill, and characteristic is valuable in its own way.

# MELATI WIJSEN

Melati Wijsen has been a passionate changemaker since 2013, when she and her sister Isabel founded the youth movement against plastic pollution, Bye Bye Plastic Bags. Through their campaign, they managed to get single-use plastic bags (alongside other plastic items) banned on their home island of Bali, and since then Melati has been an advocate of the power of young people. She's spoken at **TED** (multiple times!) and the United Nations, been featured around the world as a leading youth voice, and has founded **YOUTHTOPIA** to bring young people together and ignite their passions.

### Do you think that the relationship between plastic and the climate crisis is talked about enough?

While I feel both topics have gained a lot of awareness, there is still work to be done to show the connection between the two. I believe it is important that we understand this link in order to create a wholesome change. I think we need to learn about this in schools, we have to hear about this on the radio, on podcasts, and watch this in viral Tik Tok videos. It needs to be understood by everybody!

**You now run YOUTHTOPIA to bring together young people from around the world. Why do you think youth voices are so important?**

Youth voices are vital in creating change. We come from a generation that will not take "no" for an answer, we are resilient, and we are leading by example. I am hopeful that with our leadership, innovations, and sense of community we will create a sustainable world. At YOUTHTOPIA we believe that every young person can make a difference, but maybe not everybody knows how or where to start. So we provide tools, lessons, and bring together diverse young people from all around the world.

**You've been campaigning to help the environment since a very young age — only 12 years old! Is there any advice you'd give to your younger self that you've learned along the way?**

There are so many pieces of information that I would like to share. It's been a crazy adventure, one where you learn so much more than any textbook could ever teach you! My main three tips are:

## 1. BE CLEAR ON THE TYPE OF CHANGE YOU WANT TO CREATE.

## 2. SURROUND YOURSELF WITH A TEAM.

## 3. BE SERIOUS ABOUT CHANGE – BUT DON'T FORGET TO HAVE FUN TOO. STAY CREATIVE, STAY UNIQUE!

# Running
# A CAMPAIGN

So, you want to take your own larger action to protect the planet... **BUT WHERE DO YOU START?**

First steps are always the hardest ones to take, but here are some key steps to help you get started.

Tony Robbins, an entrepreneur who has coached many high-profile people (including four US presidents), once said that "goals are like magnets — they'll attract the things that make them come true." So, what do you want to make happen in the world?

# 1. Find your focus

Climate change is a massive issue. You can't wake up one morning and expect to stop the emission of greenhouse gases in every sector on Earth, or protect the planet from every one of climate change's effects **(at least, not straight away anyway).**

Instead, you need to find your key focus – pick one topic from this book that really interests you, and do some research into it. Importantly, this focus needs to be something that you're really passionate about. Campaigning takes time and effort, and you need to make sure that you really care about the issue you're working to address.

## 2. What's your BIG goal?

OK, so step 1 is done — you've found your focus! Now the next step is to think about how you're going to go about tackling it. What is your goal? At this point, think **BIG** — this should be a big accomplishment to work towards that would be the huge action you'd like to see take place. It should be a challenge, but not unachievable.

### 3. Lay out your steps

What changes can you start to make that will help you begin your journey? Think of your campaigning like a ladder — you've found the top rung of your **CHANGE-MAKING LADDER**, but now you need to find the ones lower down that will lead you there. This could be getting your school involved or putting together a team of young people to help you reach your target!

### 4. Get started

Onto one of the most important steps — what can you do now? What can you do tomorrow as your first step to work towards your smaller target? This step is probably one of the hardest, so make it super-easy and achievable. It could be putting up a **POSTER** at your school, or speaking to your teacher to get them involved.

## 5. Share your message

Change-making is not something to be done alone – the best change happens when it's **COLLECTIVE**, and has a greater impact than acting individually ever can! Who can you get involved to help your campaign? Is it your teacher, your parents or friends, your local community? And what tools will you use to spread this awareness?

Remember that any difference you make is important. It doesn't matter if your campaign is only a local one, encouraging your family, friends, and community to make a difference – are just as important as global ones. The most crucial thing is to do something to help the planet, no matter how small or big that might be. You have the power to make a positive difference, and it's time to use it!

# Avoiding
# FAKE NEWS

The internet can be a bit of a minefield when it comes to misleading or wrong information. Here are some tips to help you steer clear of the fake news all over the web:

## Check the source

Almost everyone is biased in some way, whether we realize it or not. This means that our personal beliefs will impact the way that we present facts or news online. However, some sources of information will be less biased than others, meaning that they will tell you what's happened with more accuracy than people who are looking to get a reaction or push a political opinion.

To check if a source is biased, look at their **"About Us"** website page, which political party (if any) they support, and whether their website address looks odd or fake.

Also keep an eye out for fact vs opinion. Opinions are people's thoughts and feelings, instead of the raw facts. While people presenting an opinion can be truthful, it's best to form your own thoughts instead of just reading someone else's.

# Give it a Google

Seen a video on a website that doesn't seem quite right? Have a look elsewhere online and see if any reliable sources (such as fact-checking websites, or scientific papers) are also sharing this news. Often, especially on social media, people will post what they think will get a reaction, as opposed to what's completely true.

## The five Ws checklist

Who, What, Where, When, Why – the five Ws! These simple words can be a great tool for avoiding fake news online. If you're not sure whether what you're reading or watching is correct, run through these words to help you out.

**WHO:** who wrote the information you're reading? Are they an expert in their field, or just someone on social media with an opinion?

**WHAT:** what is it telling you? Are there any facts, or is it just someone's thoughts?

**WHERE:** where has this information been shared? Is it a newspaper, a scientific book, or a personal blog?

**WHEN:** when was it posted? Is it still correct, or is it outdated or proven wrong?

**WHY:** why was it posted online? Is it just there to tell you something, or to push a point of view?

Did you know that a survey found that some people were more scared of public speaking than death?

Yes, public speaking is scary! But it is a useful skill to learn, especially if we're getting involved with environmental change-making.

# PUBLIC speaking

Having a good script and knowing it well is the best way to feel more confident about giving a talk. It's also the key to making your speech a really good one! So, how do you go about writing it?

## Plan your content

When it comes to public speaking, content is king!
Try making a list of the things you want to cover –
these could be **KEY FACTS** that you want to teach
people, or a campaign that you're running that you
want to tell people about.

## Outline the structure

All good talks have a thought-out beginning, middle, and end –
basically a journey that you take your listener along as you speak. You
need to make sure your content is well-ordered. If you want to teach
people about climate change, this information needs to come before
you talk about personal changes you've made, or a campaign you've
run. Otherwise, people will be sat wondering "why is this important?"

### ALSO TRY TO END WITH A CALL TO ACTION - PEOPLE SHOULD FEEL INSPIRED AND ENABLED TO MAKE A POSITIVE DIFFERENCE.

| ACTION PLAN | | | | |
|---|---|---|---|---|
| | | | | |
| | | | | |
| | | | | |
| | | | | |
| | | | | |
| | | | | |

## Plan for your audience

Tailor your message to your audience and
give them a clear goal. So, if speaking to
politicians, this could be through pushing
for government changes, or for the public
this could be calling for an action in their
everyday life.

Bear in mind your length. Us kids tend to have a shorter attention
span than adults, no matter how interesting the topic! Bring along
props to keep your classmates interested, or put something
**INTERACTIVE** into your talk to get people participating. If it's a
longer talk, try adding a short video to break up the time.

Don't worry if this seems overwhelming. The skill of writing the perfect
talk will come with time and experience, and there's no better way to find
out what works and what doesn't than by trying things out yourself!

# Presenting
# YOUR SPEECH

The way you present your talk can make or break it! In order for your presentation to be **engaging to listen to,** here are a few tips to bear in mind:

## Prepare beforehand

Did you know that you have muscles in your throat that help you to speak? To work best, these need **WARMING UP**, in the same way you'd warm up your leg muscles before running a marathon! Warming up your voice helps with your breathing and with clearing your throat, meaning we're less likely to have that moment in everyone's worst nightmare where we walk on stage and start coughing.

Here are a few warm-ups to try:

GIVE YOUR JAW A WIGGLE TO LOOSEN IT UP.

TONGUE TWISTERS (TRY OUT YOUR OWN) - FOR EXAMPLE, "RED LORRY, YELLOW LORRY".

HUM HIGHER AND LOWER - MAKE YOUR "HMMMMM" AS LONG AS IT CAN BE!

# Practise your body language

There's something called the **7-38-55** concept of communication – it sounds fancy, but basically says that only 7% of the meaning we take from speaking actually comes from the words themselves. The other 93% comes from how we act and speak – 38% is from the way we speak (our tone), and 55% is from our body language. This means that the way we act on stage is just as important as what we say!

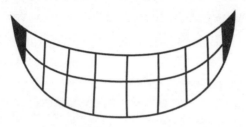

Even if you're nervous, **SMILE**! It may sound like a strange technique, but it helps to relax your body and makes you feel more confident. It also makes the audience feel a bit more relaxed! Try not to hunch or make yourself smaller on stage as well – if you keep your shoulders back and legs shoulder-width apart, it makes you appear more confident to your audience.

# Delivery

Make sure to **PACE YOURSELF** when you're delivering your talk. You always speak faster than you think, so deliver your talk at a pace that seems slow to you – the chances are, it won't feel that way to your audience who are hearing your talk for the first time! If you're not confident doing this, practise in front of your mirror, or use a voice recorder app to see how fast you're delivering your talk – this also helps to familiarize yourself with the structure.

This sounds like a lot to remember, but as soon as you're up there delivering your talk, you'll be surprised how easily it all comes to you. Enjoy yourself, because your talk will often be over before you know it!

# Writing for a
# PURPOSE

Getting your message across well on paper is a key skill to have as a campaigner. It allows you to get people on board without having to speak to them, and means you can engage with more people from places around the world.

To communicate effectively with your writing, there are a few questions to consider.

## What are you writing?

There are lots of different forms of writing to try – letters, petitions, social media posts, flyers, and more. It's important to bear in mind which one of these you're using before you get started, as they all have different structures, tones, and lengths. Some are also better for different purposes. For example, a petition is perfect for trying to get a high-ranking decision-maker to do or change something, while a flyer is a great way to get individuals to learn more about what you're doing.

## What do you want to achieve?

Be it a letter, leaflet, or petition, your writing needs to have a purpose – something that the reader can take away and do. This could be an action, or even just to learn more about climate change. Make this takeaway clear and obvious.

## Who will be reading?

**TONE** – how you, as a writer, make your reader feel – is key to getting your message across well, and needs to fit with the person you're writing to.

# Who to write to
# AND HOW

Tailoring your letter, petition, or leaflet to the correct person is super important. Different people have different levels of understanding, seniority, and abilities with which to share your message. Here are some examples of the kinds of people you could write to and things to consider:

## POLITICIANS

### Keep your message short
Politicians and their staff are very busy!

### Start locally
Try getting in touch with the politician that represents your local area, and see if they can help to spread your message.

### Include facts
Politicians want to see how offering support would help the country and local area.

### Be reasonable
It's tempting to ask for big targets, like "instantly switch all cars to electric ones!", but politicians only have so much power and time. Choose your call to action wisely. Don't be insulting, even if you don't agree with their politics!

# HEADTEACHERS

### Be polite
Use a polite, quite formal tone and carefully chosen words.

### Support your statements
Show how listening to you
would benefit your school.

### Offer solutions
Suggest an action that can be
easily taken, such as arranging a
meeting between yourself and
your headteacher.

### Offer to help drive the change
Headteachers, like politicians, are... you guessed it, busy!
Help them out by leading the way and giving your own
suggestions of what your school could do.

# BUSINESSES

### Aim for a particular person
It's best to find a particular individual to write to. This could be the
sustainability manager, or even the CEO!

### Push the idea of accountability
Show why their business should care! For
example, you may have found 100 pieces
of plastic litter with their brand on it.

### Stay realistic
Compromise by finding a small step to
take to begin with, and work from there.

### Say you can work with them
Rather than point the finger, say that you
can help them fix the problem.

# Presenting
# TO CAMERA

One of the key things that makes communicating with people successful is the way in which you bounce off your audience and their reactions. When you're making a video, this is of course much harder to do, because you can't see who'll be watching it!

Despite this, the skills of speaking to camera are pretty much like those you need when doing a talk to people in person. Here are some key things to think about when presenting to camera:

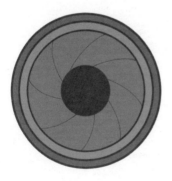

### Look straight down the lens

It might not seem noticeable to you when recording, but if you look back at your film and you're looking slightly off camera, it's pretty obvious. This is also something to bear in mind if you're filming selfie-style — don't fall into the trap of looking at the screen that shows you presenting! Make sure to look straight down the lens of the camera facing you.

## Think about your body language

If you're sitting down, try to keep a good posture, and don't be afraid to use your hands to gesture. Props are also a great way to keep your videos interesting, and you can always do a cut-shot (where you jump from one shot to the next) to zoom in on what you're holding.

## Be aware of your delivery

Vary your tone so that people don't tune out from what you're saying, and don't speak too fast so that people can't keep up. One great thing about a video is that you can always go back and watch what you've just recorded. You might be surprised how different you sound in a recording compared to how you think you sound in real life!

# Setting up your
## VIDEO

One of the most important things about making a good video – or at the very least, a watchable one – is your set-up. Having good audio and video quality makes or breaks a video, but you don't need expensive cameras or specialist microphones to achieve this.

Find a location that's quiet, with little background noise, and make sure there aren't lots of people around. Having a microphone will help reduce background interference, but it's not completely necessary.

Modern phone-cameras have pretty incredible filming abilities, which means that you don't need a super-specialist camera to film a video!

Using a tripod to hold your phone is also useful, to avoid having a shaky film that can be a bit difficult to watch. But, just as important as your camera is your composition — basically, how the elements of your film (including you!) are arranged in the frame.

There are a few different types of shots to consider when you're making a film — don't think that you have to be limited to one:

 **A close-up** is great for showing facial expression and detail, without the distraction of a background.

 **A long shot** is good for setting the scene, as it shows the background as well as you. This means that someone watching has more context about where and what you're filming. A long shot with no subject (a general view of the background only) is called an establishing shot, and is handy to use when you want to introduce the context of the film.

 **A medium shot** is from the waist up. It's a popular shot to use as it often includes the hands. This type of shot helps with capturing movement and emotion, as well as showing off your background.

 **An extreme close-up** means that the subject (what you're filming) fills the frame — or even spills out of it! This might be part of a face or a close-up of something that is important in your film.

# PLANNING
## your video

Video-making might sound super simple, but you'd be surprised how much work goes on behind the camera to make a great video! Just like with other forms of presenting your message – like writing or public speaking – the planning stage of making a video is a hugely important step. Here's how to get started planning your own change-making film.

Every minute, over 500 hours of video are uploaded onto YouTube! There are tonnes of videos on the internet, so you need to make yours stand out!

What you include in the first five seconds needs to get someone to stop scrolling and think "I want to hear what they have to say!"

## LENGTH

Think about where you want to put your video after you've made it. If you want it to be shared online, keep it short and snappy – often, people won't watch a video if it's more than one or two minutes long.

## INFORMATION

Want your video to be informational? Plan out the key facts you want to include, and put them into a structure that will have the most impact – don't just fire all the info at your audience in one go, as this will just make them feel overwhelmed and give them a bit of a headache. Instead, share some information at the start to set the scene, and then use other facts to back up your points.

## ELEVATOR PITCH

What on Earth is an elevator pitch? Well, the basic concept is: if you were taking an elevator with someone for 30 seconds, and you only had that short period of time to sell yourself and your message, what would you say? It helps with keeping your message concise and not rambling on, and also makes sure that you get across what you want to say. If you want to talk about yourself or what you're doing in your video, try giving your own elevator pitch to the camera!

This is your "hook" – something interesting that keeps your audience watching, like a quote or the beginning of a story!

# YES!

> We believe young people have the power to change the world for the better!

From the Fridays for Future strikes that millions of young people have taken part in to the amazing kids and young adults featured in this book – change can happen. Even **THROUGH YOU**, picking up this book and reading it! Young people have the voice and the potential to make a huge, positive difference. And not only is it great for the planet when kids stand up and take positive action, it's also good for ourselves. We develop skills that will help us for the rest of our lives, such as how to communicate in writing, or speak to audiences of our friends and classmates **(SOMETHING A LOT OF ADULTS STILL STRUGGLE WITH!)**.

But, for most of us kids, it can be really daunting to know where to start. How can one individual actually make a big difference on such a huge planet, and how can I actually take action? Doubts like these can knock self-belief and become barriers to action.

> Ella and I faced these barriers when we first started our own work and we don't want you to be put off by them! So, we came up with YES! – the Youth Empowerment Scheme to help young people like you get started.

**YES!** is an awards program that works to **EMPOWER** young people to take their own action by giving the support needed to get started. Through the development of key skills such as public speaking, writing for different audiences, and effective communication, YES! will help you with every step of your journey into positive action.

> If you want to learn more about the key skills needed to spread your voice far and wide, please get involved. Find YES! on: www.the-yes.co.uk

# Responding to
# FAKE NEWS

Sometimes, a false fact or misleading question can catch you off guard. Climate deniers can commonly use the same set of false information, so here are some rebuttals to help you shape your perfect answer:

## "Climate change is completely natural!"

It's true that the Earth's climate does change due to natural factors, such as volcanoes erupting. However, the current pattern of climate change just doesn't look like past natural cycles at all.

Since climate records began, scientists have not seen a period of warming so extreme or so quick. To try and prove warming is natural, climate sceptics examine short periods of time, or a specific location where the average temperature might be cooler. However, this is not evidence that climate change isn't real, because it ignores global trends.

# "How can global warming exist — it was colder than normal this winter!"

This is such a common argument against climate change that even the former US president Donald Trump used it. He said that the USA needed a bit more **"good old global warming"** after record snowfall.

It might seem confusing – how can **"warming"** cause cold weather? Well, scientists have found that the impacts of climate change on weather patterns and ocean currents can even mean we experience more cold or extreme weather regionally in the winter, alongside those scorching summers.

## "The scientists aren't sure — not all of them agree!"

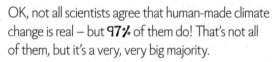

OK, not all scientists agree that human-made climate change is real – but **97%** of them do! That's not all of them, but it's a very, very big majority.

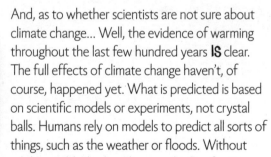

And, as to whether scientists are not sure about climate change... Well, the evidence of warming throughout the last few hundred years **IS** clear. The full effects of climate change haven't, of course, happened yet. What is predicted is based on scientific models or experiments, not crystal balls. Humans rely on models to predict all sorts of things, such as the weather or floods. Without them, we'd understand the world a lot less than we do. So, if we're going to start doubting whether we should use scientific models, we have an even bigger job to do...

# Activists
## FAQ

When speaking to people about climate change, you might get asked the same questions a lot. Here are some of the most common questions people have, and how you might want to go about answering them.

### Can I make a difference as a young person?

**OF COURSE!** Age should never stop you from standing up for something that you're passionate about. Whether it's through your own individual changes to help the planet, or through raising your voice and pushing for wider action, you can have a huge positive impact. To prevent the climate crisis from getting any worse, we all need to get involved and do our bit – old and young.

### Will my own lifestyle changes really help the planet?

Changing your lifestyle to be more sustainable can be good for the planet, and good for you! It's also so important in tackling climate change – it might not feel like making a change to your own life will make a difference, but imagine if a billion people did the same. That would have a huge positive impact on the planet.

But individual changes have to be accompanied by bigger action from government and businesses – that's why it's key that we put pressure on them to become more sustainable as well.

### How do I get started?

Start small! Begin by learning more about climate change, and telling your friends and family about the problem. Then, try and make one change in your life to help the planet – whether that's walking to school one day a week, or joining an environmental group.

## DON'T TRY AND DO EVERYTHING AT ONCE.

That will just leave you overwhelmed!

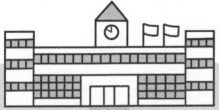

# Useful RESOURCES

Want to find out even more about climate change?
Here are a few resources that might help you out.

**Earth.Org** is an environmental organization that shares scientific understanding of climate change. And guess what – they've done a version of their information for us kids. Hurray!

**KIDS.EARTH.ORG**

**NASA** is one of the best sources when it comes to understanding climate change, and it has an amazing website for youngsters, with lots of games and activities.

**CLIMATEKIDS.NASA.GOV**

Seen an argument against climate change that makes you a bit confused? **Skeptical Science** will help you out. They break down the arguments against climate change science with different levels of difficulty, with all the graphs and models you need.

**SKEPTICALSCIENCE.COM**

### Cranky Uncle

Scientist and cartoonist John Cooke came up with this funny game that uses quizzes and illustrations to help you convince your digital cranky uncle that climate change does, in fact, exist!

**CRANKYUNCLE.COM**

Want some easy resources that you or your teacher can use to learn more about climate change? Luckily, we've compiled a whole bank of them on our charity **Kids Against Plastic's** website.

**KIDSAGAINSTPLASTIC.CO.UK/LEARN/CLIMATECHANGE**

**National Geographic Kids** have a website full of games, quizzes, videos, and even their own magazine to help understand how we can help our planet.

**KIDS.NATIONALGEOGRAPHIC.COM/GAMES/QUIZZES**

Harry from **Renewable English** (page 134) has got loads of interviews, tips, and videos from his daughter Ali to check out.

**RENEWABLEENGLISH.COM**

# GLOSSARY

## BIODEGRADABLE
something that breaks down or decays naturally

## BIODIVERSE
variety of living things in a particular area that are dependent on each other

## BIOFUEL
fuel that is made from plant or animal matter

## BOTHSIDESISM
presenting an issue as being more balanced between two opposing viewpoints, or sides, than the evidence supports

## CARBON CYCLE
circulation of carbon throughout nature

## CARBON DIOXIDE ($CO_2$)
greenhouse gas made of carbon and oxygen; it occurs naturally, but levels are rising

## CARBON FOOTPRINT
amount of carbon dioxide released into the atmosphere by the activities of one person or a group of people

## CLIMATE CHANGE
changes in the Earth's climate; this has always occurred, but is now speeding up because of global warming

## COP26
the 26th United Nations Climate Change Conference, held in 2021

**DECOMPOSITION**
process of rotting and
decay of plants and animals

**DEFORESTATION**
cutting down trees and
destroying forests

**DOOMISM**
idea that taking action to
reduce the threat of climate
change is pointless because
it's already too late

**ECO-FRIENDLY**
not harmful to the
environment

**ECOSYSTEM**
plants and animals that rely
on each other and the
environment they live in

**EMISSION**
production or release of
something, such as gas

**FLUORINATED GASES**
human-made gases
that contribute to
global warming

**FOOD SYSTEM**
interlinked activities
involved with
producing food; a farm
is a food system

**FOSSIL FUEL**
fuel, such as coal, oil, or
natural gas, made from the
long-dead remains of plants
and animals; burning fossil
fuels is one of the leading
causes of global warming

**GLOBAL WARMING**
increase in Earth's
average temperature

## GREENHOUSE GASES
gases in the Earth's atmosphere that absorb the Sun's radiation; carbon dioxide and methane are examples of greenhouse gases

## INDIGENOUS
people who originated in a particular country or place

## JET STREAM
very strong wind that has an important influence on the weather on Earth

## LANDFILL
method of burying huge amounts of rubbish

## MICROPLASTICS
tiny pieces of plastic that form when plastic objects break up in the ocean

## MONOCULTURE
large area of land that grows a single crop

## NET ZERO
achieving a balance between the greenhouse gases put into the atmosphere and those taken out

## NGO
any non-government organization that works independently from government to address a particular issue

## RENEWABLE ENERGY
form of energy that does not run out, such as wind or solar energy

## RESPIRATION
process of breathing: inhaling oxygen and exhaling carbon dioxide

## SINGLE-USE PLASTICS
plastic items, such as plastic bags, that are used only once before being thrown away or recycled

## SUSTAINABLE
practice of using natural resources at a steady level, which protects the planet now and in the future

## SUSTAINABLE DEVELOPMENT GOALS (SDGs)
17 goals to tackle global issues; the goals were devised by the UN and backed by countries around the world

## TED TALKS
online video talks posted by the American media organization TED (Technology, Entertainment, Design) Conferences LLC

## UNITED NATIONS (UN)
international organization with nearly 200 member states that aims to promote friendly relations between countries

## WATER VAPOUR
water in the form of gas; when water has just boiled it produces water vapour

# INDEX

## A few thank yous from Amy:

A huge thank you to the brilliant team at DK for bringing this book to life, and to my awesome family for their support and encouragement along the way. A thank you also to the incredible interviewees not only for sparing time in their busy lives to answer my questions, but also for their tireless work in preserving the planet for the future.

And most of all, thank you to those who are on the frontline of fighting climate change. To the climate scientists – to Mann, Bradley, and Hughes for their work on the Hockey Stick graph. To the campaigners helping to mitigate the impacts of climate change all around the world. To the young people we meet and work with who provide a much-needed burst of hope, energy, and passion to tackle the climate crisis before it's too late. And, to the Indigenous peoples and communities - those who have done the least to threaten the precious planet we live on, yet do the most to protect it.

## The publisher would like to thank:

Kieran Jones and John Hort for additional editorial help; Polly Goodman for proofreading; Helen Peters for the index; Nicola Evans for legal advice; Lauren Gardner at Bell Lomax Moreton. Most of all thank you to Amy and Ella - you've been a joy to work with throughout!

**Picture credits:**
62–3, 108–9 and 114–15: Shutterstock.com: Kingpin